SpringerBriefs in Probability and Mathematical Statistics

SpringerBriefs present concise summaries of cutting-edge research and practical applications across a wide spectrum of fields. Featuring compact volumes of 50 to 125 pages, the series covers a range of content from professional to academic. Briefs are characterized by fast, global electronic dissemination, standard publishing contracts, standardized manuscript preparation and formatting guidelines, and expedited production schedules.

Typical topics might include:

- A timely report of state-of-the art techniques
- A bridge between new research results, as published in journal articles, and a contextual literature review
- A snapshot of a hot or emerging topic
- Lecture of seminar notes making a specialist topic accessible for non-specialist readers
- SpringerBriefs in Probability and Mathematical Statistics showcase topics of current relevance in the field of probability and mathematical statistics

Manuscripts presenting new results in a classical field, new field, or an emerging topic, or bridges between new results and already published works, are encouraged. This series is intended for mathematicians and other scientists with interest in probability and mathematical statistics. All volumes published in this series undergo a thorough refereeing process.

The SBPMS series is published under the auspices of the Bernoulli Society for Mathematical Statistics and Probability.

All titles in this series are peer-reviewed to the usual standards of mathematics and its applications.

Ciprian Tudor

Non-Gaussian Selfsimilar Stochastic Processes

Springer

Ciprian Tudor
Département de Mathématiques
Université de Lille
Villeneuve-d'Ascq, France

ISSN 2365-4333   ISSN 2365-4341 (electronic)
SpringerBriefs in Probability and Mathematical Statistics
ISBN 978-3-031-33771-0   ISBN 978-3-031-33772-7 (eBook)
https://doi.org/10.1007/978-3-031-33772-7

This Springer imprint is published by the registered company Springer Nature Switzerland AG
The registered company address is: Gewerbestrasse 11, 6330 Cham, Switzerland

Preface

The origin of this short monograph lies in the lectures I gave at the Finnish Summer School on Probability and Statistics, held in Lammi, Finland, in May 2022. It concerns a particular class of self-similar stochastic processes, the so-called Hermite processes. Self-similar processes are stochastic processes that are invariant in distribution under a suitable time scaling. The most known self-similar process is the fractional Brownian motion (fBm), which can be defined as the only Gaussian self-similar process with stationary increments. Its stochastic analysis constitutes an important research direction in probability theory nowadays. The Hermite processes are non-Gaussian extensions of the fBm. These processes, which are also self-similar, with stationary increments and exhibit long-range dependence, have been also intensively studied in the last decades. The purpose is to offer a rather detailed description of this class of stochastic processes and to discuss some recent developments concerning their stochastic and statistical analysis. We analyze some stochastic (partial) differential equations driven by a Hermite process, as well as the estimation of certain parameters in stochastic models with Hermite noise.

The manuscript starts with an introduction to Wiener chaos and multiple stochastic integrals Chap. 1. This first part is motivated by the fact that almost all the random variables and processes appearing throughout this manuscript can be expressed as finite sums of multiple stochastic integrals. The main part of the material included in this introductory chapter is already contained in some references on the analysis on Wiener space or Malliavin calculus, such as [26 or 27].

In Chaps. 2–4 we introduce the Hermite processes and their multiparameter version, the Hermite sheets. We discuss their main properties, as well as the basic aspects of the stochastic integration with respect to them. We also included an analysis of some stochastic (partial) differential equations driven by an Hermite noise. In particular, we present a detailed study of the Hermite Ornstein-Uhlenbeck process (the solution of the Langevin equation with Hermite noise) and of the solution to the stochastic heat equation driven by such a random perturbation.

The last part of this book (Chap. 5) concerns the parameter estimation for Hermite-related models. While the statistical inference for stochastic equation driven by the Brownian motion, or more general, by a Gaussian noise, has a long history, the statistical inference for systems driven by Hermite processes and sheets is only at its beginning. We here present some techniques to estimate certain parameters in stochastic models involving a Hermite noise.

Although we assume that the reader has some background on the basics of probability theory and stochastic process, our intention to keep the book self-contained, as much as possible.

Villeneuve-d'Ascq, France   Ciprian Tudor

About This Book

The stochastic processes treated in this short book are mainly characterized by the following facts: they are self-similar with stationary increments and they belong to a Wiener chaos of fixed order.

The self-similarity (or the scaling property) of a object roughly means that a part of it resembles to the whole object. This property can be observed in nature or in the real life in many situations. For example, the coastlines, the top of trees or the internet traffic are approximately self-similar. For a stochastic process $(X_t, t \geq 0)$, the self-similarity can be described by the following: there exists a real-numer $H \in (0, 1)$ such that for every real number $c > 0$, the stochastic processes $(X_{ct}, t \geq 0)$ and $(c^H X_t, t \geq 0)$ have the same finite-dimensional distributions. We say that the process X is H-self-similar.

A stochastic process $(X_t, t \geq 0)$ has stationary increments if the statistical characteristics of its increments of fixed lenght do not vary over time. This property is also observed in pratice for physical phenomena driven by a stationary source. Mathematically speaking, the process $(X_t, t \geq 0)$ has stationary increments if for every $h > 0$, the finite-dimensional distributions of the stochastic processes $(X_{t+h} - X_h, t \geq 0)$ and $(X_t, t \geq 0)$ are the same.

The most proeminent example of a self-similar stochastic process with stationary increments is the fractional Brownian motion $(B_t^H, t \geq 0)$. It is defined as a zero-mean Gaussian process with covariance function

$$R(t, s) = \mathbf{E} B_t^H B_s^H = \frac{1}{2}\left(t^{2H} + s^{2H} - |t - s|^{2H}\right), \tag{1}$$

for any $s, t \geq 0$. The index H is assumed to be in the interval $(0, 1)$ and it is called the Hurst parameter, or the Hurst index. This parameter characterizes the main properties of the fractional Brownian motion (fBm) such as the regularity of the sample paths or the scaling property (the fBm is H-self-similar). If $H = \frac{1}{2}$, the fBm is nothing else than the well-known standard Brownian motion.

If a stochastic process $(X_t, t \geq 0)$ is H-self-similar and it has stationary increments, then the exact formula of its covariance can be obtained. If we assume that X is centered and X_1 has unit variance, then the covariance of the process X is given by the right-hand side of (1). Since the expectation and the covariance function determine the law of a Gaussian process, it follows that there exists (modulo a multiplicative constant) only one Gaussian self-similar process with stationary increments which is the fractional Brownian motion.

There exists a vast (recent or less recent) literature on fBm and our goal is not the focus on the analysis of this particular stochastic process. The purpose here is analyze a larger class of self-similar processes with stationary increments, the so-called Hermite processes. The class of Hermite processes includes the fractional Brownian motion but it mainly contains non-Gaussian processes. A particularity of the Hermite processes is that they live in a Wiener chaos of fixed order.

The Wiener chaos can be defined with respect to any isonormal Gaussian process defined on a probability space $(\Omega, \mathcal{F}, P)$, in particular with respect to the Brownian motion. The first Wiener chaos (i.e. the Wiener chaos of order one) contains only Gaussian random variables while the elements of the other chaoses have non-Gaussian distribution. Two Wiener chaoses of different orders are orthogonal with respect with respect to the scalar product in $L^2(\Omega)$. The Wiener chaos expansion constitutes a fundamental result in stochastic analysis. This result states that any square integrable random variable F (mesurable with respect to the sigma-algebra general by the underlying Gaussian process) can be expressed as an orthogonal infinite sum of Wiener chaoses, i.e.

$$F = \sum_{n \geq 0} F_n$$

where $F_0 = \mathbf{E}F$ and for each $n \geq 1$, the term F_n belongs to the nth Wiener chaos, denoted $\mathcal{H}_n$. In the case when the underlying Gaussian process is the standard Brownian motion, each random variable F_n can be written as a nth iterated Itô integral with respect to the Brownian motion, which is also called multiple stochastic integral (or multiple Wiener-Itô integral) of order $n \geq 1$.

The Hermite process $\left(Z_t^{H,q}, t \geq 0 \right)$ is characterized by its order $q \geq 1$ (q is an integer number) and its self-similarity index, a real number $H \in \left(\frac{1}{2}, 1 \right)$. The Hermite process of order q belongs to the qth Wiener chaos (in the sense that $Z_t^{H,q}$ belongs to $\mathcal{H}_q$ for every $t \geq 0$). When $q = 1$, the corresponding Hermite process is the fractional Brownian motion, which is the only Gaussian Hermite process. The random variable $Z_t^{H,q}$ can be then written as a multiple stochastic integral of order q with respect to the Wiener process. Therefore, we start our book with a detailed presentation of the Wiener chaos and of the multiple stochastic integrals.

After this first chapter dedicated to the construction and the properties of the multiple stochastic integrals and Wiener chaos, the rest of the material is consecrated to the analysis of the Hermite processes. We survey the findings on the distributional and trajectorial properties of this processes obtained in the past decades. As a general matter of fact, the stochastic analysis of this class of stochastic processes is challenging due to the following facts: their probability distribution is pretty complex and they are not semi-martingales. One of the purposes is to analyze the law of this processes. Some properties of this law can be obtained from its integral representation (such as the scaling property, the moments or the stationarity of the increments). As a consequence of these properties, we can also derive some of the trajectorial properties of the Hermite processes (the regularity of the sample paths or the behavior of its p-variation). But the general characterization of the distribution of a Hermite process is not known. While the covariance function of a Hermite process, of any order $q \geq 1$, il always given by (1), this covariance determines the distribution of the process only for $q = 1$ (the Gaussian case). In the case $q = 2$, the corresponding Hermite process is also called the Rosenblatt process. It was introduced in 35 and then it was made more popular by the works [16, 40] or more recently [45]. This stochastic process will receive a particular attention in our notes, because it belongs to the second Wiener chaos and in this case we know some facts about its probability distribution (which is characterized by its cumulants, or equivalently, by its moments). For a Hermite process of order $q \geq 3$, we have little knowledge about its probability law.

Besides the analysis of the basic properties of the class of Hermite processes, another goal of this manuscript is to develop a stochastic analysis with respect to these stochastic processes. As mentioned above, the main difficulty comes from the fact that the Hermite processes are not semi-martingales (they are actually zero quadratic variation processes). Therefore the classical Itô calculus cannot be applied to them. A stochastic integration theory is nowadays well-developed for the fractional Brownian motion by using various approaches (Malliavin calculus, rough paths theory, stochastic calculus via regularization etc) but these methods cannot be directly applied to the Hermite processes of order $q \geq 2$ due to their non-Gaussian character. Although there are some (pretty technical) attempts to define a stochastic integral of random integrands with respect to the Hermite processes, we will restrict here to the case of deterministic integrands, by presenting the construction and the basic properties of the Wiener integral with respect to the Hermite process. A related question considered in the manuscript is the analysis of the so-called Hermite Ornstein-Uhlenbeck, defined as the solution to the Langevin equation with Hermite noise.

We will also introduce a multiparameter version of the Hermite process, i.e. a stochastic field

$$\left(Z_t^{H,q,d}, t \in \mathbb{R}^d \right).$$

This random field is called the multiparameter Hermite process, or the Hermite sheet. We will discuss various aspects of it. This random field is also self-similar with stationary increments but these concepts has now to be understood in a multi-dimensional context. After a survey of the main properties of the multiparameter stochastic processes, we introduce a Wiener integral with respect to them. This allows to consider stochastic partial differential equations (SPDEs) with Hermite noise. We here illustrate the case of the stochastic heat equation (although other types of SPDEs, such as the stochastic wave equation, have been treated in the literature). We study when the mild solution is well-defined and we analyze the distribution and sample paths of this solution.

The last section is devoted to some applications of our theoretical results to statistics. Although the literature on statistical inference for Hermite-driven models is not yet very vast, we present two situations when various parameters appearing in such models can be estimated. These two situations concern the case of the Langevin equation with Hermite noise and of the stochastic heat equation driven by an additive Hermite sheet. In these examples, the estimation technique is based on the p-variation of the observed process, which constitutes a standars method for parameter estimation. For a given stochastic process$(X_t, t \geq 0)$, its p-variation is usually defined as

$$\sum_{i=0}^{N-1} \left| X_{t_{i+1}} - X_{t_i} \right|^p$$

for some $p > 0$, where $t_i = \frac{i}{N}$ for $i = 1, \ldots, N$. In terms of this p-variation, we define estimators for the parameters that appear in our model. By analyzing the behavior of the above sequence, we derive the asymptotic properties of the associated estimators.

Contents

Chapter 1
Multiple Stochastic Integrals

The random variables and the stochastic processes discussed below will in general be defined as multiple stochastic integrals (or multiple Wiener-Itô integrals) and they belong to a Wiener chaos. Therefore, we start our manuscript with a preliminary chapter where we included the main properties of the Wiener chaos.

The Wiener chaos is constructed with respect to the so-called isonormal Gaussian process and we start with a presentation of this concept. Then the rest of the chapter focuses on the definition and on the main properties of the Wiener chaos and of the multiple stochastic integrals.

1.1 Isonormal Processes

The isonormal processes are families of Gaussian random variables indexed by an Hilbert space. They are the basis of the construction of the Wiener chaos. We will describe them, by focusing on the examples which we are going to use in these lectures: the Brownian motion and the Brownian sheet.

Let $(\Omega, \mathcal{F}, P)$ be a probability space. Let $(H, \langle \cdot, \cdot \rangle_H)$ be a real and separable Hilbert space. Let $\| \cdot \|_H$ denote the norm in H.

Definition 1.1 A centered Gaussian family $(W(h), h \in H)$ on $(\Omega, \mathcal{F}., P)$ such that for every $h_1, h_2 \in H$,

$$\mathbf{E} W(h_1) W(h_2) = \langle h_1, h_2 \rangle_H \qquad (1.1)$$

is called an isonormal process.

It follows from Definition 1.1 that for every $h \in H$, the random variable $W(h)$ satisfies

$$W(h) \sim N\left(0, \|h\|^2\right). \qquad (1.2)$$

© The Author(s), under exclusive license to Springer Nature Switzerland AG 2023
C. Tudor, *Non-Gaussian Selfsimilar Stochastic Processes*,
SpringerBriefs in Probability and Mathematical Statistics,
https://doi.org/10.1007/978-3-031-33772-7_1

It can also be shown that the mapping $h \in H \to W(h) \in L^2(\Omega)$ is linear.

A particular case is when $H = L^2(T, \mathcal{B}, \mu)$, where T non-empty, $\mathcal{B}$ is a sigma-algebra included in $\mathcal{B}(T)$, and μ is a sigma-finite measure without atoms. In this case, we will use the notation

$$W(A) := W(1_A), \text{ for } A \in \mathcal{B}_b \tag{1.3}$$

where we denoted by $\mathcal{B}_b$ the set of $A \in \mathcal{B}$ with $\mu(A) < \infty$. The family $(W(A), A \in \mathcal{B}_b)$ satisfies the following properties:

- For every $A \in \mathcal{B}_b$, from (1.2),

$$W(A) \sim N(0, \mu(A)).$$

- For every $A \in \mathcal{B}_b$, from (1.1),

$$\mathbf{E} W(A) W(B) = \mu(A \cap B). \tag{1.4}$$

- For every $A \in \mathcal{B}_b$, the random variables $W(A)$ and $W(B)$ are independent if and only if $\mu(A \cap B) = 0$, as a consequence of (1.4) and of the fact that the family $(W(A), A \in \mathcal{B}_b)$ is Gaussian.

An easy way to construct an isonormal process in an arbitrary real and separable Hilbert space H is as follows. Let $(e_j, j \geq 1)$ be an orthonormal complete system in H and let $(X_j, j \geq 1)$ be a family of independent real-valued standard normal random variables. For every $h \in H$, set

$$W(h) = \sum_{j \geq 1} \langle h, e_j \rangle_H X_j.$$

Then $(W(h), h \in H)$ is a Gaussian family of centered random variables and for every $h_1, h_2 \in H$, we have

$$\mathbf{E} W(h_1) W(h_2) = \langle h_1, h_2 \rangle_H.$$

So $(W(h), h \in H)$ is an isonormal process.

Other useful examples of isonormal processes are given below.

1.1.1 The Wiener Integral with Respect to the Wiener Process

The two-sided Brownian motion is a centered Gaussian process $(W(t), t \in T \subset \mathbb{R})$ with covariance given by

$$\mathbf{E} W_t W_s = \frac{1}{2} (|t| + |s| - |t - s|), \quad s, t \in T. \tag{1.5}$$

When $T = [0, \infty)$, then $(W_t, t \geq 0)$ is called the standard Brownian motion. In this case, we have

$$\mathbf{E} W_t W_s = t \wedge s, \text{ for every } s, t \geq 0.$$

Remark 1.1 There is an alternative way to define the two-sided Wiener process. Consider two independent (standard) Brownian motions $(W_t^{(1)}, t \geq 0)$ and $(W_t^{(2)}, t \geq 0)$. We set $(W_t, t \in \mathbb{R})$ by

$$W_t = \begin{cases} W_t^{(1)} & \text{for} \quad t \geq 0 \\ W_{-t}^{(2)} & \text{for} \quad t < 0. \end{cases} \tag{1.6}$$

Then for every $s, t \in \mathbb{R}$,

$$\mathbf{E} W_t W_s = \begin{cases} t \wedge s, \text{ if } s, t \geq 0 \\ 0 \text{ if } t < 0, s \geq 0 \text{ or } s < 0, t \geq 0 \\ (-t) \wedge (-s) \text{ if } s, t < 0 \end{cases} = \frac{1}{2} \left(|t| + |s| - |t - s| \right). \tag{1.7}$$

Hence $(W_t, t \in \mathbb{R})$ is a centered Gaussian process with covariance (1.5).

Let $T \subset \mathbb{R}$ and let $(W_t, t \in T)$ be a Wiener process. We can naturally associate to it an isonormal process indexed by the Hilbert space $(L^2(T), \mathcal{B}, \lambda)$ where $\mathcal{B}$ is a sigma-algebra included in $\mathcal{B}(T)$ and λ is the Lebesque measure. Let $a < b$ and $A = [a, b]$. We set

$$W(A) = W_b - W_a.$$

Let $\mathcal{E}_1$ be the set of functions $h \in H = L^2(T)$ of the form

$$h(t) = \sum_{i=1}^{N} c_i 1_{A_i}(t), \quad t \in T. \tag{1.8}$$

with $N \geq 1$, $c_i \in \mathbb{R}$, and $A_1, ..., A_n$ disjoint segments in $\mathcal{B}_b$ (the set of Borel subsets with finite Lebesgue measure in $\mathcal{B}$). Then we set

$$W(h) = \sum_{i=1}^{N} c_i W(A_i).$$

We notice that

$$\mathbf{E} W(h)^2 = \sum_{i=1}^{N} c_i^2 \mathbf{E} W(A_i)^2 = \sum_{i=1}^{N} c_i^2 \lambda(A_i) = \|h\|_{L^2(T)}^2. \tag{1.9}$$

Now, let $h \in L^2(T)$. Since the set $\mathcal{E}_1$ is dense in $H = L^2(T)$, there exists a sequence $(h_n, n \geq 1)$ of simple functions of the form (1.8) such that $\|h_n - h\|_{L^2(T)} \to_{n \to \infty} 0$.

We set

$$W(h) = \lim_{n \to \infty} W(h_n) \text{ in } L^2(\Omega). \tag{1.10}$$

It is standard to see that:

- the above limit exists
- the limit (1.10) does not depend on the chosen approximating sequence $(h_n, n \geq 1)$.

Indeed, $(W(h_n), n \geq 1)$ is a Cauchy sequence in $L^2(\Omega)$ since for $m, n \geq 1$,

$$\mathbf{E}\left(W(h_m) - W(h_n)\right)^2 = \|h_m - h_n\|^2_{L^2(T)} \to_{m,n \to \infty} 0$$

since $(h_n, n \geq 1)$ converges in $L^2(T)$ (to h). Thus $(W(h_n), n \geq 1)$ converges in $L^2(\Omega)$. Also, if we consider two sequences $(h_n, n \geq 1)$ and $(g_n, n \geq 1)$ such that

$$\|h_n - h\|_{L^2(T)} \to_{n \to \infty} 0 \text{ and } \|g_n - h\|_{L^2(T)} \to_{n \to \infty} 0,$$

then

$$\mathbf{E}\left(W(h_n) - W(g_n)\right)^2 = \|h_n - g_n\|^2_{L^2(T)} \leq 2\left(\|h_n - h\|^2_{L^2(T)} + \|g - g_n\|^2_{L^2(T)}\right) \to_{n \to \infty} 0.$$

Then $(W(h), h \in H = L^2(T))$ becomes an isonormal process. Indeed, for every $h, g \in H = L^2(T)$, let $(h_n, n \geq 1)$ and $(g_n, n \geq 1)$ be two sequence of simple functions of the form (1.8) such that $\|h_n - h\|_{L^2(T)} \to_{n \to \infty} 0$ and $\|h_n - h\|_{L^2(T)} \to_{n \to \infty} 0$. Then

$$|\langle h_n, g_n \rangle_H - \langle h, g \rangle_H|$$
$$\leq |\langle h_n - h, g_n \rangle_H| + |\langle g_n - g, h \rangle_H| \leq \|h_n - h\|_H \|g_n\|_H + \|h\|_H \|g_n - g\|_H$$
$$\to_{n \to \infty} 0$$

and

$$\mathbf{E}W(h)W(g) = \lim_{n \to \infty} \mathbf{E}W(h_n)W(g_n) = \lim_{n \to \infty} \langle h_n, g_n \rangle_H = \langle h, g \rangle_H.$$

We also use the notation

$$W(h) = \int_T h(s) dW_s.$$

The random variable $W(h)$ is called the Wiener integral of h with respect to the Wiener process.

Remark 1.2 Conversely, to any isonormal process indexed by a Hilbert space of L^2-type, one can associate a Brownian motion. Let $(W(h), h \in L^2(\mathbb{R}_+))$ be an isonormal process. Set, for every $t \geq 0$,

$$B_t = W\left(1_{[0,t]}\right).$$

Then $(B_t, t \geq 0)$ is a centered Gaussian process, starting from zero and for every $s, t \geq 0$,

$$\mathbf{E}B_t B_s = \mathbf{E}W\left(1_{[0,t]}\right) W\left(1_{[0,s]}\right) = \langle 1_{[0,t]}, 1_{[0,s]} \rangle_{L^2(\mathbb{R}_+)} = t \wedge s.$$

Consequently, $(B_t, t \geq 0)$ is a standard Brownian motion.

1.1.2 The Wiener Integral with Respect to the Brownian Sheet

The Brownian sheet constitutes a multiparameter version of the process defined by (1.5).

Definition 1.2 Let $N \geq 1$. The N-parameter Brownian sheet is defined as a centered Gaussian process $(W(x), x \in T \subset \mathbb{R}^N)$ with covariance

$$\mathbf{E}W(x)W(y) = \prod_{j=1}^{N} \left(\frac{1}{2}(|x_j| + |y_j| - |x_j - y_j|) \right), \qquad (1.11)$$

if $x = (x_1, ..., x_N), y = (y_1, ..., y_N) \in T \subset \mathbb{R}^N$.

If $N = 1$, then (1.11) reduces to (1.5). If $T = \mathbb{R}^N_+$, then the process with covariance (1.11) will be called the standard Brownian sheet. So, in this situation,

$$\mathbf{E}W(x)W(y) = \prod_{j=1}^{N} (x_j \wedge y_j).$$

We also introduce the notion of the high-order increment of a d- parameter process $X = (X(x), x \in \mathbb{R}^d)$ on a rectangle $[s, t] \subset \mathbb{R}^d, s = (s_1, ..., s_d), t = (t_1, ..., t_d)$, with $s \leq t$. This increment is denoted by $\Delta X_{[s,t]}$ and it is given by

$$\Delta X_{[s,t]} = \sum_{r \in \{0,1\}^d} (-1)^{d - \sum_i r_i} X_{s + \mathbf{r} \cdot (t - s)}. \qquad (1.12)$$

When $d = 1$ one obtains the

$$\Delta X_{[s,t]} = X_t - X_s$$

while for $d = 2$ one gets

$$\Delta X_{[s,t]} = X_{t_1,t_2} - X_{t_1,s_2} - X_{s_1,t_2} + X_{s_1,s_2}.$$

We can associate an isonormal process with the Wiener sheet $(W(x), x \in T \subset \mathbb{R}^N)$. If A is a rectangle of the form $A = [a, b] = [a_1, b_1] \times [a_2, b_2] \times \ldots \times [a_N, b_N]$, then we set

$$W(A) = \Delta W_{[a,b]}.$$

The definition can be then extended to simple functions of the form

$$h(x) = \sum_{i=1}^{M} c_i 1_{A_i}(x), x \in \mathbb{R}^N,$$

where $M \geq 1$, $c_i \in \mathbb{R}$ and A_i are disjoint rectangles in $\mathcal{B}_b(T)$, for $i = 1, \ldots, M$, by setting

$$W(h) = \sum_{i=1}^{M} c_i W(A_i).$$

By the density of the simple functions in $L^2(T)$, we can extend $W(h)$, for every $h \in L^2(\mathbb{R}^N)$ by (1.10).

By following the one-parameter case, we can show that the family $(W(h), h \in L^2(T)$ becomes an isonormal process, in the sense of Definition 1.1, i.e. it is a centered Gaussian process such that

$$\mathbf{E}W(h_1)W(h_2) = \langle h_1, h_2 \rangle_{L^2(\mathbb{R}^N)} = \int_{\mathbb{R}^N} h_1(x_1, \ldots, x_N) h_2(x_1, \ldots, x_N) dx_1 \ldots dx_N.$$

We also use the notation

$$W(h) = \int_{T^N} h(x) dW(x)$$

and we will call $W(h)$ the Wiener integral of h with respect to the Wiener sheet W.

1.2   Multiple Wiener-Itô Integrals

1.2.1   Definition and Basic Properties

We will present the construction of the Wiener chaos and of the multiple stochastic integrals on a Hilbert space of L^2-type, although the construction of these objects can be done over a general underlying Hilbert space (see e.g. [27] or [29]). Let $H = L^2(T, \mathcal{B}, \mu)$ and let $(W(h), h \in H)$ an isonormal process. We denote by $\mathcal{E}_n$ the set of elementary functions of n variables $f : T^n \to \mathbb{R}$. We say that $f \in \mathcal{E}_n$ if f can be written as

$$f(t_1, \ldots, t_n) = \sum_{i_1,\ldots,i_n=1}^{N} a_{i_1,\ldots,i_n} 1_{A_{i_1} \times A_{i_2} \times \ldots \times A_{i_n}}(t_1, \ldots, t_n), \qquad (1.13)$$

where $N \geq 1$, the set $A_1, \ldots, A_N \in \mathcal{B}_b$ are disjoint and $a_{i_1,\ldots,i_n} = 0$ if two indices are equal (i.e. if there exist $k \neq l$ with $i_k = i_l$).

For instance $f(t_1, t_2) = 21_{[0,1) \times [1,2)}(t_1, t_2)$ belongs to $\mathcal{E}_2$ but $f(t_1, t_2, t_3) = 21_{[0,1) \times [1,2) \times [0,1)}(t_1, t_2, t_3)$ does not belong to $\mathcal{E}_3$.

We first define the multiple stochastic integral of an elementary function. Recall the notation (1.3).

Definition 1.3 For $f : T^n \to \mathbb{R}$ of the form (1.13), we set

$$I_n(f) = \sum_{i_1,\ldots,i_n=1}^{N} a_{i_1,\ldots,i_n} W(A_{i_1}) \ldots W(A_{i_n}) \qquad (1.14)$$

We will call $I_n(f)$ the multiple stochastic integral of order n of f with respect to isonormal process W. Notice that $I_n(f)$ belongs to $L^2(\Omega)$.

If $f : T^m \to \mathbb{R}$, we denote by $\tilde{f}$ its symmetrization, i.e.

$$\tilde{f}(t_1, \ldots, t_m) = \frac{1}{n!} \sum_{\sigma \in S_n} f(t_{\sigma_1}, \ldots, t_{\sigma_m}). \qquad (1.15)$$

We have the following lemma.

Lemma 1.1 For every $f \in L^2(T^n)$,

$$\|\tilde{f}\|_{L^2(T^n)} \leq \|f\|_{L^2(T^n)}.$$

Proof We have, by (1.15) and Cauchy-Schwarz,

$$\|\tilde{f}\|_{L^2(T^n)} = \frac{1}{n!^2} \int_{T^n} \left(\sum_{\sigma \in S_n} f(t_{\sigma(1)}, \ldots, t_{\sigma(n)}) \right)^2 dt_1 \ldots dt_n$$

$$\leq \frac{1}{n!} \sum_{\sigma \in S_n} \int_{T^n} f^2(t_{\sigma(1)}, \ldots, t_{\sigma(n)}) dt_1 \ldots dt_n$$

$$= \frac{1}{n!} \sum_{\sigma \in S_n} \|f\|_{L^2(T^n)}^2 = \|f\|_{L^2(T^n)}^2.$$

∎

Proposition 1.1 *For $m \geq 1$, the mapping*

$$I_m : \mathcal{E}_m \to L^2(\Omega)$$

satisfies the following properties.

1. *$I_m : \mathcal{E}_m \to L^2(\Omega)$ is linear.*
2. *We have*

$$I_m(f) = I_m(\tilde{f})$$

   where $\tilde{f}$ denotes the symmetrization of f defined by (1.15).
3. *We have the "isometry" of multiple integrals: for $f \in \mathcal{E}_p, g \in \mathcal{E}_q$,*

$$\mathbf{E}\left(I_p(f)I_q(g)\right) = \begin{cases} q! \langle \tilde{f}, \tilde{g} \rangle_{L^2(T^p)} & \text{if } p = q \\ 0 & \text{otherwise.} \end{cases} \tag{1.16}$$

In particular, for $f \in \mathcal{E}_p$,

$$\mathbf{E}I_p(f)^2 = p! \|\tilde{f}\|_{L^2(T^p)}.$$

Proof For 1., we can always assume that $f, g \in \mathcal{E}_m$ are expressed in terms of the same partition, then it is trivial from the definition (1.14) that

$$I_m(\alpha f + g) = \alpha I_m(f) + I_m(g).$$

To prove 2., we can assume, by linearity, that

$$f(t_1, ..., t_m) = 1_{A_1 \times .... \times A_m}(t_1, ..., t_m)$$

with $A_1, ..., A_m$ disjoint sets in $\mathcal{B}_b$. Then

$$\tilde{f}(t_1, ..., t_m) = \frac{1}{m!} \sum_{\sigma \in S_m} 1_{A_{\sigma(1)} \times ... \times A_{\sigma(m)}}(t_1, ..., t_m)$$

and by (1.14),

$$\begin{aligned} I_m(\tilde{f}) &= \frac{1}{m!} \sum_{\sigma \in S_m} W(A_{\sigma(1)}).....W(A_{\sigma(m)}) \\ &= \frac{1}{m!} \sum_{\sigma \in S_m} W(A_1)....W(A_m) \\ &= W(A_1).....W(A_m) = I_m(f). \end{aligned}$$

We now prove 3. Let $f \in L^2(T^m)$ of the form (1.13) and let $g \in L^2(T^m)$,

$$g(t_1, .., t_m) = \sum_{j_1,...,j_m=1}^{N} b_{j_1,...,j_m} 1_{A_{j_1} \times .... \times A_{j_m}}(t_1, ..., t_m).$$

Then,

$$\mathbf{E} I_n(f) I_m(g)$$
$$= \sum_{i_1,...,i_n,j_1,...,j_m=1}^{N} a_{i_1,...,i_n} b_{j_1,...,j_m} \mathbf{E} \left(W(a_{i_1}) \ldots W(A_{i_n}) W(A_{j_1}) \ldots W(A_{j_m}) \right).$$

Assume $m \neq n$, say $m > n$. Then there always exists $k = 1, ..., m$ with $j_k \notin \{i_1, ..., i_m\}$ and

$$\mathbf{E} \left(W(a_{i_1}) \ldots W(A_{i_n}) W(A_{j_1}) \ldots W(A_{j_m}) \right) = \mathbf{E} (\ldots) \mathbf{E}(W(A_{j_k}) = 0.$$

Let $m = n$. Then

$$\mathbf{E} I_n(f) I_n(g)$$
$$= \mathbf{E} \left(n! \sum_{i_1 < ... < i_n} a_{i_1,...,i_n} W(A_{i_1}) .... W(A_{i_n}) \right) \left(n! \sum_{j_1 < ... < j_n} b_{j_1,...,j_n} W(A_{j_1}) .... W(A_{j_n}) \right)$$
$$= (n!)^2 \sum_{i_1 < ... < i_n, j_1 < .. < j_n} a_{i_1,...,i_n} b_{j_1,...,j_n} \mathbf{E} \left(W(A_{i_1}) .... W(A_{i_n}) W(A_{j_1}) .... W(A_{j_n}) \right).$$

But $\mathbf{E} \left(W(A_{i_1}) .... W(A_{i_n}) W(A_{j_1}) .... W(A_{j_n}) \right) = 0$ if $(i_1, ..., i_n) \neq (j_1, ..., j_n)$. So,

$$\mathbf{E} I_n(f) I_n(g)$$
$$= (n!)^2 \sum_{i_1 < ... < i_n} a_{i_1,...,i_n} b_{i_1,...,i_n} \mathbf{E} W(A_{a_i})^2 \ldots \mathbf{E} W(A_{i_n})^2$$
$$= (n!)^2 \sum_{i_1 < ... < i_n} a_{i_1,...,i_n} b_{i_1,...,i_n} \mu(A_{i_1}) \ldots \mu(A_{i_n}) = n! \langle \tilde{f}, \tilde{g} \rangle_{L^2(T^n)}.$$

∎

The Definition 1.3 of the multiple Wiener-Itô integral can be extended to any $f \in L^2(T^n)$, $n \geq 1$. It can be shown that the set $\mathcal{E}_n$ is dense in $L^2(T^n)$ and thus for every $f \in L^2(T^n)$ there exists a sequence $(f_k, k \geq 1) \subset \mathcal{E}_n$ such that

$$\| f_k - f \|_{L^2(T^n)} \to_{k \to \infty} 0.$$

Then we define

$$I_n(f) = L^2(\Omega) - \lim_{k \to \infty} I_n(f_k). \tag{1.17}$$

Notice that the limit in (1.17) exists. Indeed, for $k, l \geq 1$, by Lemma 1.1 and the isometry property (1.16),

$$\mathbf{E} \left(I_n(f_k) - I_n(f_l) \right)^2 = \mathbf{E} \left(I_n(f_k - f_l) \right)^2$$
$$= n! \| \widetilde{f_k - f_l} \|^2_{L^2(T^n)} \leq n! \| f_k - f_l \|^2_{L^2(T^n)} \to_{k,l \to \infty} 0$$

because $(f_k, k \geq 1)$ is convergent, and thus a Cauchy sequence, in $L^2(T^n)$.

Also, the limit (1.17) does not depend on the approximating sequence. Take two sequences $(f_k, k \geq 1)$ and $(g_k, k \geq 1)$ such that

$$\| f_k - f \|_{L^2(T^n)} \to_{k \to \infty} 0 \text{ and } \| g_k - f \|_{L^2(T^n)} \to_{k \to \infty} 0.$$

Then, as above,

$$\mathbf{E} \left(I_n(f_k) - I_n(g_k) \right)^2 \leq n! \| f_k - g_k \|^2_{L^2(T^n)}$$
$$\leq 2n! \left(\| f_k - f \|^2_{L^2(T^n)} + \| f - g_k \|^2_{L^2(T^n)} \right) \to_{k \to \infty} 0.$$

Consequently, $(I_n(f_k), k \geq 1)$ and $(I_n(g_k), k \geq 1)$ have the same limit in $L^2(\Omega)$.

The extension of I_n to $L^2(T^n)$ satisfies the same properties as for simple functions (Proposition 1.1).

Proposition 1.2 • *The application* $I_n : L^2(T^n) \to L^2(\Omega)$ *is linear.*
• *For every* $f \in L^2(T^n)$, *we have* $I_n(f) = I_n(\tilde{f})$, *with* $\tilde{f}$ *given by (1.15).*
• *The property (1.16) holds true for every* $f \in L^2(T^n), g \in L^2(T^m)$.

Proof Point 1. is trivial by using the linearity of I_n on simple functions and taking then the limit. For point 2., if $(f_k, k \geq 1) \subset \mathcal{E}_n$, then $I_n(f_k) \to_{k \to \infty} I_n(f)$ in $L^2(\Omega)$ and

$$\mathbf{E} \left(I_n(\tilde{f_k}) - I_n(\tilde{f}) \right)^2 = n! \| \widetilde{f_k - f} \|^2_{L^2(T^n)} \leq n! \| f_k - f \|^2_{L^2(T^n)} \to_{k \to \infty} 0.$$

By Proposition 1.1, point 2.,

$$I_n(f_k) = I_n(\tilde{f_k}) \text{ for every } k \geq 1$$

and we take the limit as $k \to \infty$ in the above relation.

Concerning 3., the case $m \neq n$ follows easily by a limit argument while for $m = n$, since

$$I_n(f_k)I_n(g_k) - I_n(f)I_n(g) = I_n(f_k) \left(I_n(g_k) - I_n(g) \right) + I_n(g) \left(I_n(f_k) - I_n(f) \right),$$

we get

$$\mathbf{E} |I_n(f_k) I_n(g_k) - I_n(f) I_n(g)|$$
$$\leq \mathbf{E} |I_n(f_k) (I_n(g_k) - I_n(g))| + \mathbf{E} |I_n(g) (I_n(f_k) - I_n(f))|$$
$$\leq \left(\mathbf{E} I_n(f_k)^2\right)^{\frac{1}{2}} (\mathbf{E}(I_n(g_k) - I_n(g)))^{\frac{1}{2}} + \left(\mathbf{E} I_n(g)^2\right)^{\frac{1}{2}} (\mathbf{E}(I_n(f_k) - I_n(f)))^{\frac{1}{2}} \to_{k\to\infty} 0$$

and, by writing

$$\langle \tilde{f}_k, \tilde{g}_k \rangle_{L^2(T^n)} - \langle \tilde{f}, \tilde{g} \rangle_{L^2(T^n)} = \langle \tilde{f}_k - \tilde{f}, \tilde{g}_k \rangle_{L^2(T^n)} + \langle \tilde{f}, \tilde{g}_k - \tilde{g} \rangle_{L^2(T^n)}$$

we will have

$$\left| \langle \tilde{f}_k, \tilde{g}_k \rangle_{L^2(T^n)} - \langle \tilde{f}, \tilde{g} \rangle_{L^2(T^n)} \right|$$
$$\leq \left| \langle \tilde{f}_k - \tilde{f}, \tilde{g}_k \rangle_{L^2(T^n)} \right| + \left| \langle \tilde{f}, \tilde{g}_k - \tilde{g} \rangle_{L^2(T^n)} \right|$$
$$\leq \| f_k - f \|_{L^2(T^n)} \| g_k \|_{L^2(T^n)} + \| f \|_{L^2(T^n)} \| g_k - g \|_{L^2(T^n)}$$

and this goes to zero as $k \to \infty$. ∎

Remark 1.3 If $m = 1$ and $T \subset \mathbb{R}^N$, then the multiple integral I_1 coincides with the Wiener integral defined in Sects. 1.1.1 and 1.1.2.

1.2.2 A First Product Formula

The product formula for multiple stochastic integrals will be intensively used in our work. This formula says how the product of two multiple integral $I_m(f) I_n(g)$ can be expressed as a sum of multiple stochastic integrals. We start with the case $n = 1$.

Let $L_s^2(T^n)$ be the set of symmetric functions in $L^2(T^n)$. If $f \in L_s^2(T^m)$ and $g \in L_s^2(T^n)$, we define the contraction of order r ($0 \leq r \leq m \wedge n$) of f and g by

$$(f \otimes_r g)(t_1, ..., t_{m+n-2r})$$
$$= \int_{T^r} f(t_1, ..., t_{m-r}, u_1, ..., u_r) g(t_{m-r+1}, ..., t_{m+n-2r}, u_1, ..., u_r) du_1 ... du_r. \quad (1.18)$$

If $r = 0$, then

$$f \otimes_0 g = f \otimes g,$$

the tensor product of f and g.

Lemma 1.2 Let $f \in L^2(T^m)$ and $g \in L^2(T^n)$ be two symmetric functions. Then

$$f \otimes_r g \in L^2(T^{m+n-2r}) \text{ for } r = 0, 1, ..., m \wedge n.$$

Proof By (1.18), for $r = 1, ..., m \wedge n,$

$$\|f \otimes_r g\|^2_{L^2(T^{m+n-2r})}$$

$$= \int_{T^{m+n-2r}} dt_1 \ldots dt_{m+n-2r}$$

$$\left(\int_{T^r} f(t_1, ..., t_{m-r}, u_1, ..., u_r) g(t_{m-r+1}, ..., t_{m+n-2r}, u_1, ..., u_r) du_1 ... du_r \right)^2$$

$$\leq \int_{T^{m+n-2r}} dt_1 \ldots dt_{m+n-2r} \left(\int_{T^r} f^2(t_1, ..., t_{m-r}, u_1, ..., u_r) du_1 ... du_r \right)$$

$$\times \left(\int_{T^r} g^2(t_{m-r+1}, ..., t_{m+n-2r}, v_1, ..., v_r) dv_1 ... dv_r \right)$$

where we used Cauchy-Schwarz's inequality. So,

$$\|f \otimes_r g\|^2_{L^2(T^{m+n-2r})}$$

$$\leq \left(\int_{T^m} dt_1 ... dt_{m-r} du_1 ... du_r f^2(t_1, ..., t_{m-r}, u_1, ..., u_r) \right)$$

$$\times \left(\int_{T^n} dt_{m-r+1} ... dt_{m+n-2r} dv_1 ... dv_r g^2(t_{m-r+1}, ..., t_{m+n-2r}, v_1, ..., v_r) dv_1 ... dv_r \right)$$

$$= \|f\|^2_{L^2(T^m)} \|g\|^2_{L^2(T^n)}.$$

The result is also true for $r = 0$.   ∎

Remark 1.4 We notice that in general the contraction $f \otimes_r g$ is not a symmetric function even if f and g are symmetric,. Indeed, take

$$f = 1_{\widetilde{[1,2] \times [3,4]}} \text{ and } g = 1_{[1,2] \times [1,2]}.$$

Then

$$(f \otimes_1 g)(t_1, t_2) = 1_{[1,2] \times [3,4]}(t_2, t_1) \text{ and } (f \otimes_1 g)(t_2, t_1) = 1_{[1,2] \times [3,4]}(t_1, t_2).$$

Let us state and prove a first version of the product formula for multiple stochastic integrals.

Proposition 1.3 Let $f \in L^2_s(T^m)$ and $g \in L^2(T)$. Then

$$I_m(f)I_1(g) = I_{m+1}(f \otimes g) + m I_{m-1}(f \otimes_1 g) \tag{1.19}$$

where $f \otimes_1 g$ is the 1-contraction of f and g given by (1.18).

Proof Let us first assume that

$$f = 1_{\widetilde{A_1 \times ... \times A_m}} = \frac{1}{m!} \sum_{\sigma \in S_m} 1_{A_{\sigma(1)} \times .... \times A_{\sigma(m)}}.$$

with $A_1, ..., A_m$ disjoint sets in $\mathcal{B}_b$. Let $g = 1_B$ with $B \in \mathcal{B}_b$.

If $A_1, ..., A_m, B$ are mutually disjoint, then $f \otimes g$ is a simple function in $\mathcal{E}_{m+1}$ with $f \otimes_1 g = 0$. Then

$$I_m(f)I_1(g) = W(A_1) ... W(A_m)W(B) = I_{n+1}(f \otimes g).$$

Assume $g = 1_B$ where $B \cap A_1 = C_1 \neq \emptyset$ and $B, A_2, ..., A_m$ disjoint. We write

$$A_1 \setminus B = D_1, B_0 = B \setminus C_1.$$

For given $\varepsilon > 0$, we decompose C_1 as

$$C_1 = \bigcup_{k=1}^N E_k \text{ with } \mu(E_k) < \varepsilon \text{ and } E_k \text{ disjoints for } k = 1, ..., N.$$

Then

$$
\begin{aligned}
&I_n(f)I_1(g) \\
&= W(A_1) .... W(A_m)W(B) \\
&= (W(C_1) + W(D_1))W(A_2) .... W(A_m)(W(C_1) + W(B_0)) \\
&= W(C_1)^2 W(A_2) .... W(A_m) \\
&\quad + (W(C_1)W(B_0) + W(C_1)W(D_1) + W(D_1)W(B_0))\, W(A_2) .... W(A_m).
\end{aligned}
$$

Now, we write

$$W(C_1)^2 = \left(\sum_{k=1}^N W(E_k)\right)^2 = \sum_{k=1}^N W(E_k)^2 + \sum_{k,l=1;k\neq l}^N W(E_k)W(E_l).$$

So,

$$
\begin{aligned}
&I_m(f)I_1(g) \\
&= \sum_{k=1}^N W(E_k)^2 W(A_2) .... W(A_m) + \sum_{k,l=1;k\neq l}^N W(E_k)W(E_l)W(A_2) .... W(A_m) \\
&\quad + (W(C_1)W(B_0) + W(C_1)W(D_1) + W(D_1)W(B_0))\, W(A_2) .... W(A_m). \quad (1.20)
\end{aligned}
$$

Set

$$h_\varepsilon = \sum_{k,l=1; k\neq l}^{N} 1_{E_k \times E_l \times A_2 \times .... \times A_m}$$
$$+ 1_{C_1 \times B_0 \times A_2 \times ... \times A_m} + 1_{C_1 \times D_1 \times A_2 \times ... \times A_m} + 1_{D_1 \times B_0 \times A_2 \times ... \times A_m}.$$

By (1.20),

$$I_m(f)I_1(g)$$
$$= I_{m+1}(h_\varepsilon) + \sum_{k=1}^{N} W(E_k)^2 W(A_2)....W(A_m)$$
$$= I_{m+1}(h_\varepsilon) + \sum_{k=1}^{N} \left(W(E_k)^2 - \mu(E_k) \right) W(A_2)....W(A_m)$$
$$+ \sum_{k=1}^{N} \mu(E_k) W(A_2)....W(A_m)$$
$$= I_{m+1}(h_\varepsilon) + R_\varepsilon + \mu(C_1) W(A_2)....W(A_m)$$

with

$$R_\varepsilon = \sum_{k=1}^{N} \left(W(E_k)^2 - \mu(E_k) \right) W(A_2)....W(A_m).$$

Next, we observe that, by (1.18),

$$(f \otimes_1 g)(t_1, ..., t_{m-1})$$
$$= \int_T \frac{1}{m!} \sum_{\sigma \in S_m} 1_{A_{\sigma(1)} \times .... \times A_{\sigma(m)}}(t_1, ..., t_{m-1}, u) 1_B(u) du$$
$$= \frac{1}{m!} \sum_{\sigma \in S_m} 1_{A_{\sigma(1)} \times .... \times A_{\sigma(m-1)}}(t_1, ..., t_{m-1}) \int_T 1_{A(\sigma(m)}(u)(1_{C_1}(u) + 1_{B_0}(u)) du$$
$$= \frac{1}{m!} \sum_{\sigma \in S_m} 1_{A_{\sigma(1)} \times .... \times A_{\sigma(m-1)}}(t_1, ..., t_{m-1}) \mu(C_1) 1_{\sigma(m)=1}$$
$$= \frac{1}{m} 1_{\overbrace{A_2 \times ... \times A_m}}(t_1, ..., t_{m-1}) \mu(C_1).$$

We obtained,

$$I_m(f)I_1(g) = I_{m+1}(h_\varepsilon) + m I_{m-1}(f \otimes_1 g) + R_\varepsilon. \qquad (1.21)$$

Let us show that

$$R_\varepsilon \to_{\varepsilon \to 0} 0 \text{ in } L^2(\Omega).$$

Since $(A_2, ..., A_m, E_k, k \geq 1)$ are disjoint, then $(W(A_2), ..., W(A_m), W(E_k), k \geq 1)$ is a family of independent Gaussian random variables. Thus,

$$\mathbf{E}(R_\varepsilon^2) = \mathbf{E}\left(\sum_{k=1}^{N} \left(W(E_k)^2 - \mu(E_k)\right)\right)^2 \mathbf{E}W(A_2)^2....\mathbf{E}W(A_m)^2$$

$$= \mu(A_2)....\mu(A_m) \sum_{k=1}^{N} \mathbf{E}\left(W(E_k)^2 - \mu(E_k)\right)^2$$

$$= 2\mu(A_2)....\mu(A_m) \sum_{k=1}^{N} \mu(E_k)^2$$

$$\leq 2\varepsilon\mu(A_2)....\mu(A_m) \sum_{k=1}^{N} \mu(E_k) = 2\varepsilon\mu(A_2)....\mu(A_m) \sum_{k=1}^{N} \mu(C_1).$$

Let us now prove that $\tilde{h}_\varepsilon$ converges to $f\widetilde{\otimes}g$ in $L^2(T^{m+1})$. We have

$$\|\tilde{h}_\varepsilon - f\widetilde{\otimes}g\|_{L^2(T^{m+1})}^2 \leq \|h_\varepsilon - h\|_{L^2(T^{m+1})}^2,$$

with

$$h = 1_{C_1 \times C_1 \times A_2 \times...\times A_m} + 1_{C_1 \times B_0 \times A_2 \times...\times A_m} + 1_{C_1 \times D_1 \times A_2 \times...\times A_m} + 1_{D_1 \times B_0 \times A_2 \times...\times A_m}.$$

Therefore

$$\|\tilde{h}_\varepsilon - f\widetilde{\otimes}g\|_{L^2(T^{m+1})}^2$$

$$\leq \|\sum_{k,l=1;k\neq l}^{N} 1_{E_k \times E_l \times A_2 \times...\times A_m} - 1_{C_1 \times C_1 \times A_2 \times...\times A_m}\|_{L^2(T^{m+1})}^2$$

$$= \sum_{k=1}^{N} \mu(E_k)^2 \mu(A_2)...\mu(A_m) \leq \varepsilon\mu(C_1)\mu(A_2)...\mu(A_m).$$

By taking the limit as $\varepsilon \to 0$ in (1.21) and using the linearity of multiple stochastic integrals, we obtain, for every $f \in \mathcal{E}_m, g \in \mathcal{E}$,

$$I_m(f)I_1(g) = I_{m+1}(f \otimes g) + mI_{m-1}(f \otimes_1 g). \tag{1.22}$$

Consider now $f \in L_S^2(T^m)$ and $g \in L^2(T)$ such that

$$\|f_k - f\|_{L^2(T^m)} \to_{k \to \infty} 0 \text{ and } \|g_k - g\|_{L^2(T)} \to_{k \to \infty} 0.$$

Then clearly

$$I_m(f_k)I_1(g_k) \to_{k \to \infty} I_m(f)I_1(g) \text{ in } L^1(\Omega)$$

and

$$\mathbf{E}\,|I_{m+1}(f_k \otimes g_k) - I_{m+1}(f \otimes g)|^2$$
$$= (m+1)!\|f_k \tilde{\otimes} g_k - f \tilde{\otimes} g\|^2_{L^2(T^{m+1})} \le (m+1)!\|f_k \tilde{\otimes} g_k - f \otimes g\|^2_{L^2(T^{m+1})}$$

and since

$$f_k \otimes g_k - f \otimes g = f_k \otimes (g_k - g) + (f_k - f) \otimes g,$$

we get

$$\mathbf{E}\,\left|I_{m+1}(f_k \otimes g_k) - I_{m+1}(f \otimes g)\right|^2$$
$$\le 2(m+1)!\|f_k \otimes (g_k - g)\|^2_{L^2(T^{m+1})} + 2(m+1)!\|(f_k - f) \otimes g\|^2_{L^2(T^{m+1})}$$
$$\le 2(m+1)!\|f_k\|^2_{L^2(T^m)}\|g_k - g\|^2_{L^2(T)} + 2(m+1)!\|f_k - f\|^2_{L^2(T^m)}\|g\|_{L^2(T)} \to_{k \to \infty} 0.$$

Similarly, by writting

$$f_k \otimes_1 g_k - f \otimes g = f_k \otimes_1 (g_k - g) + (f_k - f) \otimes_1 g,$$

we will have

$$\mathbf{E}\,\left|I_{m-1}(f_k \otimes_1 g_k) - I_{m-1}(f \otimes_1 g)\right|^2$$
$$\le 2(m-1)!\|f_k\|^2_{L^2(T^m)}\|g_k - g\|^2_{L^2(T)} + 2(m-1)!\|f_k - f\|^2_{L^2(T^m)}\|g\|_{L^2(T)} \to_{k \to \infty} 0.$$

$\blacksquare$

1.2.3 The Wiener Chaos

The definition of the Wiener chaos is related to the Hermite polynomials. The Hermite polynomial of degree n is defined by $H_0(x) = 1$ for every $x \in \mathbb{R}$ and for $n \ge 1$,

$$H_n(x) = \frac{(-1)^n}{n!} e^{\frac{x^2}{2}} \frac{d^n}{dx^n} \left(e^{-\frac{x^2}{2}}\right). \tag{1.23}$$

The first Hermite polynomials are $H_1(x) = x$, $H_2(x) = \frac{1}{2}(x^2 - 1)$.

Proposition 1.4 *Consider the function* $F(x, t) = e^{tx - \frac{t^2}{2}}$ *for* $x, t \in \mathbb{R}$. *Then*

$$F(x, t) = \sum_{n \geq 0} t^n H_n(x). \tag{1.24}$$

Proof It follows from the Taylor expansion around the origin of the function $t \to F(x, t)$. Indeed, for $x, t \in \mathbb{R}$,

$$F(x, t) = 1 + \sum_{n \geq 1} \frac{\partial^n F}{\partial t^n}\Big|_{t=0} \frac{t^n}{n!},$$

and one can show that forever $n \geq 1$, $\frac{\partial^n F}{\partial t^n}\Big|_{t=0} = n! H_n(x)$.   ∎

We list the main properties of the Hermite polynomials:

Proposition 1.5 *1. For every* $n \geq 1$, *we have*

$$H_n'(x) = H_{n-1}(x) \tag{1.25}$$

2. For every $n \geq 1$, *we have*

$$(n + 1) H_{n+1}(x) = x H_n(x) - H_{n-1}(x), \quad x \in \mathbb{R} \tag{1.26}$$

3. For every $n \geq 1$,

$$H_n(-x) = (-1)^n H_n(x), n \geq 1. \tag{1.27}$$

Proof For the first property, we differentiate ith respect to x in (1.24) and we identify the coefficients of the power series. To prove 2., we differentiate with respect to t in (1.24) and we identify the coefficients of the power series. For 3., we use $F(-x, t) = F(x, -t)$.   ∎

Definition 1.4 Let $m \geq 0$. The Wiener chaos of order m (with respect to an isonormal process W) is defined as the vector subspace of $L^2(\Omega)$ generated by the random variables

$$(H_m(W(h)), h \in H, \|h\|_H = 1).$$

By definition, since $H_0(x) = 1$ for every $x \in \mathbb{R}$, the chaos of order zero is $\mathcal{H}_0 = \mathbb{R}$. The chaos of order one coincides with the Gaussian space generated by $(W(h), h \in H, \|h\|_H = 1)$. Consequently, $\mathcal{H}_1$ contains only Gaussian random variables. For $m \geq 2$, all the non-trivial elements of $\mathcal{H}_m$ are non-Gaussian random variables.

The below result gives the link between the Wiener chaos and the multiple stochastic integrals.

Proposition 1.6 *Let* $(W(h), h \in H)$ *be an isonormal process. Let* $h \in H$ *with* $\|h\|_H = 1$. *Then for every* $m \geq 1$,

$$m! H_m(W(h)) = I_m(h^{\otimes m}). \tag{1.28}$$

Proof The relation (1.28) will be proven by induction on $m \geq 1$. For $m = 1$, (1.28) holds true because

$$H_1(W(h)) = W(h) = I_1(h)$$

by the construction of multiple integrals. Assume that (1.28) holds for $1, 2, ..m$. We have, by the product formula,

$$
\begin{aligned}
I_m(h^{\otimes m}) I_1(h) &= I_{m+1}(h^{\otimes(m+1)}) + m I_{m-1}(h^{\otimes m} \otimes_1 h) \\
&= I_{m+1}(h^{\otimes(m+1)}) + m I_{m-1}(h^{\otimes(m-1)}),
\end{aligned} \tag{1.29}
$$

due to the fact that

$$h^{\otimes m} \otimes_1 h = \|h\|_H^2 h^{\otimes(m-1)} = h^{\otimes(m-1)}.$$

Using the induction hypothesis, we can write by (1.29),

$$
\begin{aligned}
I_{m+1}(h^{\otimes(m+1)}) &= m! H_m(W(h)) W(h) - m(m-1)! H_{m-1}(W(h)) \\
&= m! \left(H_m(W(h)) W(h) - H_{m-1}(W(h)) \right) \\
&= (m+1)! H_{m+1}(W(h))
\end{aligned}
$$

where we used (1.26) for the last identity.   ∎

Corollary 1 *The Wiener chaos* $\mathcal{H}_m$ *coincides with the image of* $L_S^2(T^m)$ *through the application* I_m.

Proof The first observation is that for every $m \geq 1$, and for every $h \in H$,

$$H_m(W(h)) \in I_m\left(L_S^2(T^m)\right),$$

due to Proposition 1.6, so any linear combination $\sum_{i=1}^N \lambda_i H_m(W(h_i))$ (with $\lambda_i \in \mathbb{R}, h_i \in H$) belongs to $I_m\left(L_S^2(T^m)\right)$, which is a vector space. By the isometry formula (1.16) with $m = n$, we notice the $I_m\left(L_S^2(T^m)\right)$ is also a closed subspace of $L^2(\Omega)$. Therefore

$$\mathcal{H}_m \subset I_m\left(L_S^2(T^m)\right). \tag{1.30}$$

On the other hand, by the orthogonality of multiple stochastic integrals, given $F \in I_m\left(L_S^2(T^m)\right)$, we have

$$\mathbf{E} F I_n(h^{\otimes n}) = 0, \text{ for } n \geq 1, n \neq m.$$

This, together with Proposition 1.6, implies that

$$\mathbf{E}FG = 0 \text{ for every } G \in \mathcal{H}_n, n \geq 1, n \neq m.$$

So, $I_m\left(L_S^2(T^m)\right)$ is orthogonal to any $\mathcal{H}_n$, with $n \neq m$. By (1.30), one obtains

$$\mathcal{H}_m = I_m\left(L_S^2(T^m)\right).$$

∎

Remark 1.5 It follows from Corollary 1 and by the isometry (1.16) that two Wiener chaoses of different order are orthogonal, i.e. if $F \in \mathcal{H}_m$ and $G \in \mathcal{H}_m$ with $m \neq n$, then $\mathbf{E}FG = 0$.

The next results gives the equivalence of all norm of the Wiener chaos. It is also known as the *hypercontractivity property*. For the proof, see Corollary 2.8.14 in [27].

Proposition 1.7 *Let F be an element in a fixed sum of Wiener chaoses, i.e. $F \in \bigoplus_{n=0}^{N}\mathcal{H}_n$. For for every $1 < p < q$, we have*

$$\|F\|_{L^p(\Omega)} \leq \|F\|_{L^q(\Omega)} \leq C(p,q)\|F\|_{L^p(\Omega)}.$$

In particular, for every $p > 2$,

$$\mathbf{E}|F|^p \leq C(p)\left(\mathbf{E}|F|^2\right)^{\frac{p}{2}}. \tag{1.31}$$

Remark 1.6 For further use, let us note that ijn the particular case $p = 2$, the following inequality holds: for $q > 2$,

$$\|F\|_{L^q(\Omega)} \leq (q-1)\|F\|_{L^2(\Omega)}. \tag{1.32}$$

The below result is a fundamental result in the analysis on Wiener space. We refer to [29] for its proof.

Theorem 1.1 (Wiener chaos expansion) *Any random variable $F \in L^2(\Omega, \mathcal{G}, P)$ ($\mathcal{G}$ is the sigma-algebra generated by W) can be expanded into a series of multiple stochastic integrals*

$$F = \sum_{m \geq 0} I_m(f_m),$$

with $\mathbf{E}F = I_0(f_0)$ and $f_m \in L^2(T^m)$. If the kernels f_m are assumed to be symmetric, then they are uniquely determined by F.

1.2.4   The General Product Formula

Let us now state the general product formula.

Theorem 1.2 *Let $f \in L_S^2(T^m)$ and $g \in L_S^2(T^n)$ with $m, n \geq 1$. Then*

$$I_m(f)I_n(g) = \sum_{r=0}^{m \wedge n} r! \binom{m}{r}\binom{n}{r} I_{m+n-2r}(f \otimes_r g). \tag{1.33}$$

Proof Let $m \geq n$. We prove formula (1.33) by induction on $m \geq 1$. For $m = 1$, the result has been proven in Proposition 1.3. Assume $g = g_1 \tilde{\otimes} g_2$ with $g_1 \in L_S^2(T^n)$, $g_2 \in L^2(T)$ with $g_1 \otimes_1 g_2 = 0$. Then, by (1.3), $I_n(g) = I_{n-1}(g_1)I_1(g_2)$ and by using the induction hypothesis,

$$I_m(f)I_n(g) = I_m(f)I_{n-1}(g_1)I_1(g_2)$$
$$= \sum_{r=0}^{n-1} r! \binom{m}{r}\binom{n-1}{r} I_{m+n-1-2r}(f \otimes_r g_1)I_1(g_2).$$

Now, by (1.19),

$$I_{m+n-1-2r}(f \otimes_r g_1)I_1(g_2)$$
$$= I_{m+n-1-2r}(f \tilde{\otimes}_r g_1)I_1(g_2)$$
$$= I_{m+n-2r}\left((f \tilde{\otimes}_r g_1) \otimes g_2\right) + (m+n-2r-1)I_{m+n-2r-2}\left((f \tilde{\otimes}_r g_1) \otimes_1 g_2\right)$$

Therefore

$$I_m(f)I_n(g)$$
$$= \sum_{r=0}^{n-1} r! \binom{m}{r}\binom{n-1}{r} I_{m+n-2r}\left((f \tilde{\otimes}_r g_1) \otimes g_2\right)$$
$$+ \sum_{r=0}^{n-1} r! \binom{m}{r}\binom{n-1}{r}(m+n-2r-1)I_{m+n-2r-2}\left((f \tilde{\otimes}_r g_1) \otimes_1 g_2\right)$$
$$= \sum_{r=0}^{n-1} r! \binom{m}{r}\binom{n-1}{r} I_{m+n-2r}\left((f \tilde{\otimes}_r g_1) \otimes g_2\right)$$
$$+ \sum_{r=1}^{n} (r-1)! \binom{m}{r-1}\binom{n-1}{r-1}(m+n-2r+1)I_{m+n-2r}\left((f \tilde{\otimes}_{r-1} g_1) \otimes_1 g_2\right)$$

We use the combinatorial identity

$$\frac{r(m+n-2r+1)}{m-r-1}\left((f \tilde{\otimes}_{r-1} g_1) \otimes_1 g_2\right) + (n-r)\left((f \tilde{\otimes}_r g_1) \otimes g_2\right) = n(f \tilde{\otimes}_r g_1)$$

to obtain (1.33) for $g = g_1 \widetilde{\otimes} g_2$ with $g_1 \in L_S^2(T^n), g_2 \in L^2(T)$ with $g_1 \otimes_1 g_2 = 0$.

∎

In particular, if $m = n = 2$ we get for $f, g \in L_S^2(T^2)$,

$$I_2(f)I_2(g) = I_4(f \otimes g) + 4I_2(f \otimes_1 g) + 2\langle f, g \rangle_{L^2(T^2)}$$

and if $h, g \in L^2(T)$ with $\langle h, g \rangle_{L^2(T)} = 0$, then

$$I_m(h^{\otimes m})I_n(g^{\otimes n}) = I_{m+n}(h^{\otimes m} \otimes g^{\otimes n}).$$

1.3 Random Variables in the Second Wiener Chaos

The random variables in the second Wiener chaos constitutes a special class among the multiple stochastic integrals. We have some knowledge about their probability distribution. Although they are non-Gaussian, we will show below that their probability law is completely determined by their cumulants (or, equivalently, by their moments). Moreover, it is possible to obtain an explicit formula for these cumulants.

For $f \in H^{\odot 2}$ (i.e. $f \in H^{\otimes 2}$ and f is symmetric), consider the operator

$$A_f : H \to H, \quad A_f(g) = f \otimes_1 g.$$

A classical result in functional analysis says that A_f is an Hilbert-Schmidt operator. Denote by $(\lambda_{j,f}, j \geq 1), (e_{j,f}, j \geq 1)$ the eigenvalues and the eigenvectors of A_f. Then we have

$$f = \sum_{j \geq 1} \lambda_{j,f} e_{j,f} \otimes e_{j,f}. \tag{1.34}$$

Moreover for every $p \geq 2$, we have

$$\sum_{j=1}^{\infty} \lambda_{j,f}^p < \infty$$

and

$$Tr(A_f^p) = \langle f \otimes_1^{(p-1)} f, f \rangle_{H^2} = \sum_{j=1}^{\infty} \lambda_{j,f}^p, \tag{1.35}$$

where $Tr(A_f)$ stands for the trace of the operator $A_f^p = \underbrace{A_f \circ \ldots A_f}_{p}$. The sequence $(f \otimes_1^{(m)} f, m \geq 1)$ is defined recursively: $f \otimes_1^{(1)} f = f$ and for $p \geq 2$,

$$f \otimes_1^{(p)} f = (f \otimes_1^{(p-1)} f) \otimes_1 f.$$

We deduce that the random variables have a particular form.

Proposition 1.8 *Let $f \in H^{\otimes 2}$, f symmetric. Then*

$$I_2(f) = \sum_{j \geq 1} \lambda_{j,f} \left(Z_j^2 - 1 \right) \tag{1.36}$$

where $(Z_j, j \geq 1)$ is a family of independent standard normal random variables and $\lambda_{j,f}$ are defined by (1.34).

Proof By (1.34),

$$I_2(f) = \sum_{j \geq 1} \lambda_{j,f} I_2(e_{j,f} \otimes e_{j,f}) = \sum_{j \geq 1} \lambda_{j,f} \left(I_1(e_{j,f})^2 - 1 \right).$$

It suffices to observe that $(Z_j = I_1(e_{j,f}), j \geq 1)$ is a family of i.i.d. $N(0, 1)$ random variables.   ∎

Let us denote by $k_m(F)$, $m \geq 1$ the mth cumulant of a random variable F. It is defined as

$$k_m(F) = (-i)^m \frac{\partial^m}{\partial t^m} \ln \mathbf{E}(e^{itF})|_{t=0},$$

if $F \in L^m(\Omega)$. In particular, $k_1(F) = \mathbf{E}F$ and $k_2(F) = Var(F)$. In other words, for every $u \in \mathbb{R}$,

$$\log \mathbf{E}(e^{iuF}) = \sum_{m=1}^{\infty} \frac{(iu)^m}{m!} k_m(F). \tag{1.37}$$

We have the following link between the moments and the cumulants of F: for every $m \geq 1$,

$$k_m(F) = \sum_{\sigma=(a_1,...,a_r) \in \mathcal{P}(\{1,...,n\})} (-1)^{r-1}(r-1)! \mathbf{E}X^{|a_1|} \ldots \mathbf{E}X^{|a_r|} \tag{1.38}$$

if $F \in L^m(\Omega)$, where $\mathcal{P}(b)$ is the set of all partitions of b. In particular, for centered random variables F, we have $k_1(F) = \mathbf{E}F, k_2(F) = \mathbf{E}F^2, k_3(F) = \mathbf{E}F^3, k_4(F) = \mathbf{E}F^4 - 3(\mathbf{E}F^2)^2$.

We have an explicit expression of the cumulants of the random variables in the second Wiener chaos.

Theorem 1.3 *Let $F = I_2(f)$ with $f \in L_S^2(T^2)$, Then for every $m \geq 2$,*

$$k_m(F) = 2^{m-1}(m-1)! \int_{T^m} f(u_1, u_2) f(u_2, u_3) \ldots f(u_{m-1}, u_m) f(u_m, u_1) du_1 \ldots du_m. \tag{1.39}$$

Proof The proof is taken from [27]. First, from (1.36), we can compute the characteristic function of F, obtaining, for $u \in \mathbb{R}$,

$$\mathbf{E}(e^{iuF}) = \prod_{j=1}^{\infty} \frac{e^{-iu\lambda_{j,f}}}{\sqrt{1 - 2iu\lambda_{j,f}}}.$$

By applying the logarithm above, we find

$$\log \mathbf{E}(e^{iuF}) = \sum_{m=2}^{\infty} 2^{m-1} \frac{(iu)^m}{m} \sum_{j=1}^{\infty} \lambda_{j,f}^p. \tag{1.40}$$

By identifying the coefficients in (1.37) and (1.40), we get $k_1(F) = 0$ and for $m \geq 2$,

$$k_m(F) = 2^{m-1}(m-1)! \sum_{j\geq 1} \lambda_{j,f}^m = 2^{m-1}(m-1)! \langle f \otimes_1^{(m-1)} f, f \rangle_{H^{\otimes 2}}$$

where $H = L^2(T)$ and the last equality comes from (1.35). It then remains to check that for $m \geq 2$,

$$\langle f \otimes_1^{(m-1)} f, f \rangle_{H^{\otimes 2}} = \int_{T^m} f(x_1, x_2) f(x_2, x_3) \ldots f(x_{m-1}, x_m) f(x_m, x_1) dx_1 \ldots dx_m. \tag{1.41}$$

To this end, we prove first by induction that for $m \geq 2$,

$$(f \otimes_1^{(m)} f)(x_1, x_2) = \int_{T^{m-1}} du_1 \ldots du_{m-1} f(x_1, u_1) f(u_1, u_2) \ldots f(u_{m-2}, u_{m-1}) f(u_{m-1}, x_2). \tag{1.42}$$

Formula (1.42) holds for $m = 2$, due to (1.18). Then

$$(f \otimes_1^{(m+1)} f)(x_1, x_2) = \left((f \otimes_1^{(m)} f) \otimes_1 f \right)(x_1, x_2)$$

$$= \int_T du_m (f \otimes_1^{(m)} f)(x_1, u_m) f(x_2, u_m)$$

$$= \int_T du_m f(x_2, u_m) \int_{T^{m-1}} du_1 \ldots du_{m-1} f(x_1, u_1) f(u_1, u_2) \ldots f(u_{m-1}, u_m)$$

$$= \int_{T^m} du_1 \ldots du_m f(x_1), u_1) f(u_2, u_3) \ldots f(u_{m-1}, u_m) f(u_m, x_2).$$

This implies (1.42). It then suffices to use (1.42) to obtain (1.41) and the conclusion. ∎

Finally, let us show that the cumulants (or the moments, see relation (1.38)) determines the law of a random variable in the second Wiener chaos. We say that the the law of a random variable $Y \subset \bigcap_{m\geq 1} L^m(\Omega)$ is determined by the moments (or by the cumulants) if for any other random variable $X \in \bigcap_{m \geq 1} L^m(\Omega)$ with $\mathbf{E} Y^m = \mathbf{E} X^m$ for every $m \geq 1$, we have that the law of Y coincides with the law of X.

Proposition 1.9 *Assume $f \in H^{\odot 2}$ and let $F = I_2(f)$. Then the law of F is entirely determined by the sequence of cumulants (1.39).*

Proof We will give the sketch of the proof. Actually it suffices to show that (see Exercise 2.7.12 in [27])

$$\mathbf{E} e^{t|X|} < \infty \text{ for some } t > 0.$$

By assuming $\mathbf{E} F^2 = 1$, we have by hypercontractivity (inequality (1.32)),

$$\left(\mathbf{E} |F|^p \right)^{\frac{1}{p}} \leq p - 1$$

for every $p > 2$. So, by using Markov's inequality,

$$P(|F| > u) \leq u^{-p}(p-1)^p$$

for any $u > 0$. We write now $p = 1 + \frac{u}{e}$ with $0 < e < u$. Then, via Fubini,

$$\mathbf{E}(e^{t|F|}) = 1 + t \int_0^\infty e^{tu} P(|F| > u) \, du < \infty$$

for $0 \leq t < \frac{1}{e}$.   ∎

Chapter 2
Hermite Processes: Definition and Basic Properties

Historically, the Hermite processes appeared as limit of partial sums of sequences of correlated random variables in the so-called Non-Central Limit Theorem, see e.g. [17, 41, 43]. Let us briefly recall the basic facts. Let $Z \sim N(0, 1)$ and consider a function $g : \mathbb{R} \to \mathbb{R}$ such that $\mathbf{E}g(Z) = 0$ and $\mathbf{E}g(Z)^2 < \infty$. Then g can be expanded into a basis of Hermite polynomial, i.e.

$$g(x) = \sum_{j=0}^{\infty} c_j H_j(x), \quad x \in \mathbb{R},$$

where H_j denotes the jth Hermite polynomial given by (1.23) and $c_j = \mathbf{E}(g(Z)H_j(Z))$. *The Hermite rank of g* is defined by

$$k = \min\{l \ |c_l \neq 0\}.$$

Since $\mathbf{E}[g(Z)] = 0$, we have $k \geq 1$. Assume g has Hermite rank equal to q and let $(\xi_n, n \in \mathbb{Z})$ be a stationary Gaussian sequence with mean 0 and variance 1 which exhibits long range dependence in the sense that the correlation function satisfies

$$r(n) := \mathbf{E}(\xi_0 \xi_n) = n^{\frac{2H-2}{q}} L(n)$$

where $H \in (\frac{1}{2}, 1)$, $q \geq 1$ and L is a slowly varying function at infinity (see e.g. [18] for the definition). Consider the partial sums, for $n \geq 1, t \geq 0$,

$$X_n(t) = \frac{1}{n^H} \sum_{j=1}^{[nt]} g(\xi_j). \tag{2.1}$$

Due to the relatively strong correlation of the summands of the above sum, the sequence $(X_n, n \geq 1)$ will have, in many situations, a non-Gaussian limit. Actually, the family of stochastic processes $(X_n, n \geq 1)$ converges as $n \to \infty$, in the sense

© The Author(s), under exclusive license to Springer Nature Switzerland AG 2023
C. Tudor, *Non-Gaussian Selfsimilar Stochastic Processes*,
SpringerBriefs in Probability and Mathematical Statistics,
https://doi.org/10.1007/978-3-031-33772-7_2

of finite-dimensional distributions, to a stochastic process which lives in the Wiener chaos of order q. This limit process is the Hermite process. Below, we will define it properly and we will analyze its main properties.

The class of Hermite processes contains the well-known fractional Brownin motion, which is the only Gaussian Hermite process. We will see that the Hermite process (of general order) shares many properties with fBm: covariance structure, self-similarity, stationarity of the increments, regularity of sample paths. Therefore, the Hermite process constitutes an interesting alternative to fBm for modelling purposes, especially in applications where the empirical data shows a non-Gaussian character. An example of such an application in hidrology has been provided in [42]. More recently, the Hermite processes found applications in network traffic (see e.g. [9]) or mathematical finance (see e.g. [19, 40]). Notice that a generalized version of the Hermite process (whose self-similarity index may be less than one half) has been studied in e.g. [7] or [3].

Let us describe the contain of this chapter. In the sequel we will denote by $\left(Z_t^{H,q}, t \geq 0\right)$ the Hermite process of order $q \geq 1$ with Hurst parameter (or self-similarity index) $H \in \left(\frac{1}{2}, 1\right)$.

• We start with an analysis of the kernel of the Hermite process. Being an element of the qth Wiener chaos, the random variable $Z_t^{H,q}$ can be expressed as a multiple integral of order q, i.e. $Z_t^{H,q} = I_q(L_t^{H,q})$ for every $t \geq 0$, according to Corollary 1. The function $L_t^{H,q}$ is defined on $\mathbb{R}^q$ and it is called the kernel of the Hermite process. The properties of this kernel play an important role for the stochastic analysis of the Hermite process.
• We then study the main properties of the Hermite process such as the scaling property, the stationary of its increments, the moments, the regularity of the sample paths or the p-variation.
• We included a paragraph consecrated to some particular Hermite processes: the fractional Brownian motion (which is the Hermite process of order 1 and the only Gaussian Hermite process) and the Rosenblatt process (the Hermite process of order q). For these particular case, it is possible to completely describe their finite-dimensional probability distributions by using their cumulants.
• Other integral representations of the Hermite process are also stated and proven. Actually, the multiple integral representation $Z_t^{H,q} = I_q(L_t^{H,q})$ is not unique. We show that the stochastic process $\left(Z_t^{H,q}, t \in [0, T]\right)$ coincides, in the sense of finite-dimensional distributions, with $\left(I_q(A_t^{H,q}), t \in [0, T]\right)$, where $A_t^{H,q}$ is a kernel in $L^2([0, T]^q)$ different from $L_t^{H,q}$. Depending of the problem considered, one or another representation may be more useful.
• A short discussion on the possibility to simulate the Hermite processes is included in the last part of this chapter.

2.1 The Kernel of the Hermite Process

We start by introducing the kernel of the Hermite process. For $a, b > 0$, we denote by $\beta(a, b)$ the beta function

$$\beta(a, b) = \int_0^1 x^{a-1}(1-x)^{b-1}dx. \tag{2.2}$$

Let us start with the following lemmas which will be intensively used in the sequel.

Lemma 2.1 Let $H \in \left(\frac{1}{2}, 1\right)$. Then for every $s, t \geq 0$,

$$H(2H-1)\int_0^t \int_0^s |u-v|^{2H-2}dudv = \frac{1}{2}\left(t^{2H} + s^{2H} - |t-s|^{2H}\right).$$

Proof The formula can be obtained by differentiation both sides $\frac{\partial^2}{\partial s \partial t}$. ∎

Lemma 2.2 Let $a, b, u, v \in \mathbb{R}$ such that $a + b + 1 < 0$ and $u \neq v$. Then

$$\int_{-\infty}^{u \wedge v} (u-y)^a(v-y)^b dy = \begin{cases} \beta(-1-a-b, a+1)(v-u)^{a+b+1}, & \text{if } u < v \\ \beta(-1-a-b, b+1)(u-v)^{a+b+1}, & \text{if } v < u. \end{cases}$$

In particular, if $a = b < -\frac{1}{2}$ and $u \neq v$,

$$\int_{-\infty}^{u \wedge v} (u-y)^a(v-y)^a dy = \beta(-1-2a, a+1)|u-v|^{2a+1}. \tag{2.3}$$

Proof We recall the alternative definition of the beta function: if $x, y > 0$, then

$$\beta(x, y) = \int_0^\infty t^{x-1}(1+t)^{-x-y}dt.$$

Let $u < v$. By the change of variables $w = \frac{u-y}{v-u}$, we get

$$\int_{-\infty}^u (u-y)^a(v-b)^b dy = (v-u)^{a+b+1} \int_0^\infty w^a(1+w)^b$$
$$= \beta(a+1, -1-a-b)(v-u)^{a+b+1}.$$

∎

For $t \geq 0$, we set

$$L_t^{H,q}(y_1, \ldots, y_q) = c(H, q) \int_0^t (u-y_1)_+^{-\left(\frac{1}{2}+\frac{1-H}{q}\right)} \ldots (u-y_q)_+^{-\left(\frac{1}{2}+\frac{1-H}{q}\right)} du, \tag{2.4}$$

for every $y_1, \ldots, y_q \in \mathbb{R}$, with

$$c(H, q)^2 = \frac{H(2H-1)}{q! \beta \left(\frac{1}{2} - \frac{1-H}{q}, \frac{2-2H}{q} \right)^q}. \tag{2.5}$$

Proposition 2.1 *For $t \geq 0$, let $L_t^{H,q}$ be given by (2.4). Then $L_t^{H,q} \in L^2(\mathbb{R}^q)$ and for every $s, t \geq 0$,*

$$\langle L_t^{H,q}, L_s^{H,q} \rangle_{L^2(\mathbb{R}^q)} = q! \frac{1}{2} \left(t^{2H} + s^{2H} - |t-s|^{2H} \right).$$

In particular, for every $t \geq 0$,

$$\|L_t^{H,q}\|_{L^2(\mathbb{R}^q)}^2 = q! t^{2H}.$$

Proof We have, for $s, t \geq 0$, by Fubini,

$$\langle L_t^{H,q}, L_s^{H,q} \rangle_{L^2(\mathbb{R}^q)} = \int_{\mathbb{R}^q} L_t^{H,q}(y_1, \ldots, y_q) L_s^{H,q}(y_1, \ldots, y_q) dy_1 \ldots dy_q$$

$$= c(H, q)^2 \int_{\mathbb{R}^q} dy_1 \ldots dy_q \left(\int_0^t (u - y_1)_+^{-\left(\frac{1}{2} + \frac{1-H}{q}\right)} \ldots (u - y_q)_+^{-\left(\frac{1}{2} + \frac{1-H}{q}\right)} du \right)$$

$$\times \left(\int_0^s (v - y_1)_+^{-\left(\frac{1}{2} + \frac{1-H}{q}\right)} \ldots (v - y_q)_+^{-\left(\frac{1}{2} + \frac{1-H}{q}\right)} dv \right)$$

$$= c(H, q)^2 \int_0^t du \int_0^s dv \left(\int_{\mathbb{R}} (u - y)_+^{-\left(\frac{1}{2} + \frac{1-H}{q}\right)} (v - y)_+^{-\left(\frac{1}{2} + \frac{1-H}{q}\right)} dy \right)^q.$$

We apply (2.3) with $a = -\frac{1}{2} + \frac{H-1}{q} < -\frac{1}{2}$ to get

$$\int_{\mathbb{R}} (u - y)_+^{-\left(\frac{1}{2} + \frac{1-H}{q}\right)} (v - y)_+^{-\left(\frac{1}{2} + \frac{1-H}{q}\right)} dy$$

$$= \int_{-\infty}^{u \wedge v} (u - y)_+^{-\left(\frac{1}{2} + \frac{1-H}{q}\right)} (v - y)_+^{-\left(\frac{1}{2} + \frac{1-H}{q}\right)} dy$$

$$= \beta \left(\frac{1}{2} - \frac{1-H}{q}, \frac{2-2H}{q} \right) |u - v|^{\frac{2H-2}{q}}.$$

Thus

$$
\langle L_t^{H,q}, L_s^{H,q}\rangle_{L^2(\mathbb{R}^q)} = c(H,q)^2 \beta\left(\frac{1}{2} - \frac{1-H}{q}, \frac{2-2H}{q}\right)^q \int_0^t du \int_0^s dv |u-v|^{2H-2}
$$

$$
= c(H,q)^2 \beta\left(\frac{1}{2} - \frac{1-H}{q}, \frac{2H-2}{q}\right)^q \frac{1}{H(2H-1)}\frac{1}{2}\left(t^{2H} + s^{2H} - |t-s|^{2H}\right)
$$

$$
= \frac{1}{q!}\frac{1}{2}\left(t^{2H} + s^{2H} - |t-s|^{2H}\right),
$$

where we used (2.5).   ∎

Also notice that the kernel (2.4) satisfies the following scaling property

$$
L_{ct}^{H,q}(y_1, \dots, y_q) = c^{H - \frac{q}{2}} L_t^{H,q}\left(\frac{y_1}{c}, \dots, \frac{y_q}{c}\right),
$$

for every $c > 0$ and for every $y_1, \dots, y_q \in \mathbb{R}$. This property plays a role in the proof of the self-similarity of the Hermite process.

2.2 Definition of the Hermite Process and Some Immediate Properties

We start with the definition of the Hermite process.

Definition 2.1 The Hermite process of order $q \geq 1$ and with self-similarity index $H \in \left(\frac{1}{2}, 1\right)$ is defined as $Z^{H,q} = \left(Z_t^{H,q}, t \geq 0\right)$, with

$$
Z_t^{H,q}
$$

$$
= c(H,q) \int_{\mathbb{R}^q} \left(\int_0^t (u-y_1)_+^{-\left(\frac{1}{2} + \frac{1-H}{q}\right)} \dots (u-y_q)_+^{-\left(\frac{1}{2} + \frac{1-H}{q}\right)} du\right) dB(y_1)\dots dB(y_q)
$$

$$
= I_q(L_t^{H,q}), \quad t \geq 0, \tag{2.6}
$$

where $B = (B(y), y \in \mathbb{R})$ is a Brownian motion over the whole real line (or the two-sided Wiener process, see (1.5)), I_q is the multiple integral of order q with respect to B defined in Sect. 1.2 and the kernel $L^{H,q}$ is defined by (2.4).

Notice that for every $t \geq 0$, the random variable $Z_t^{H,q}$ is well-defined and belongs to $L^2(\Omega)$, since (see Proposition 2.1)

$$
\mathbf{E}(Z_t^{H,q})^2 = q! \| L_t^{H,q} \|_{L^2(\mathbb{R}^q)}^2 = t^{2H}.
$$

Moreover, for every $s, t \geq 0$, the covariance of the process $Z^{H,q}$ reads as,

$$\mathbf{E} Z_t^{H,q} Z_s^{H,q} = R^H(t, s) := \frac{1}{2} \left(t^{2H} + s^{2H} - |t - s|^{2H} \right). \tag{2.7}$$

A Hermite random variable will be in the sequel a random variable (with index H and order q) with the same distribution as $Z_1^{H,q}$.

2.2.1 Self-similarity and Stationarity of the Increments

We will say that a stochastic process $(X_t, t \geq 0)$ is self-similar of index $H \in (0, 1)$ (or H-self-similar) if for every $c > 0$, the stochastic processes

$$(X_{ct}, t \geq 0) \text{ and } (c^H X_t, t \geq 0)$$

have the same finite-dimensional distributions.

A stochastic process $(X_t, t \geq 0)$ is said to be with stationary increments if for every $h > 0$, the stochastic processes

$$(X_t, t \geq 0) \text{ and } (X_{t+h} - X_h, t \geq 0)$$

have the same finite-dimensional distributions. We will check these properties for the Hermite process.

Proposition 2.2 *The process $Z^{H,q}$ given by (2.6) is self-similar of index H and it has stationary increments.*

Proof Let $c > 0$. Then for every $t \geq 0$,

$$Z_t^{H,q} = c(H, q) \int_{\mathbb{R}^q} \left(\int_0^{ct} (u - y_1)_+^{-\left(\frac{1}{2} + \frac{1-H}{q}\right)} \cdots (u - y_q)_+^{-\left(\frac{1}{2} + \frac{1-H}{q}\right)} du \right)$$
$$\times dB(y_1) \ldots dB(y_q)$$

$$= c \times c(H, q) \int_{\mathbb{R}^q} \left(\int_0^t (cu - y_1)_+^{-\left(\frac{1}{2} + \frac{1-H}{q}\right)} \cdots (cu - y_q)_+^{-\left(\frac{1}{2} + \frac{1-H}{q}\right)} du \right)$$
$$\times dB(y_1) \ldots dB(y_q)$$

$$= c \times c(H, q) \int_{\mathbb{R}^q} \left(\int_0^t (cu - cy_1)_+^{-\left(\frac{1}{2} + \frac{1-H}{q}\right)} \cdots (cu - cy_q)_+^{-\left(\frac{1}{2} + \frac{1-H}{q}\right)} du \right)$$
$$\times dB(cy_1) \ldots dB(cy_q)$$

$$= c^{-\frac{q}{2} + H} c(H, q) \int_{\mathbb{R}^q} \left(\int_0^t (u - y_1)_+^{-\left(\frac{1}{2} + \frac{1-H}{q}\right)} \cdots (u - y_q)_+^{-\left(\frac{1}{2} + \frac{1-H}{q}\right)} du \right)$$
$$\times dB(cy_1) \ldots dB(cy_q).$$

Since the Brownian motion B is $\frac{1}{2}$-self-similar, we have (we denote by $\equiv^{(d)}$ the equivalence of finite-dimensional distributions)

$$\left(\int_{\mathbb{R}^q} \left(\int_0^t (cu - cy_1)_+^{-\left(\frac{1}{2}+\frac{1-H}{q}\right)} \ldots (u - y_q)_+^{-\left(\frac{1}{2}+\frac{1-H}{q}\right)} du \right) \right.$$

$$\left. \times dB(cy_1) \ldots dB(cy_q), t \geq 0 \right)$$

$$\equiv^{(d)} c^{\frac{q}{2}} \left(\int_{\mathbb{R}^q} \left(\int_0^t (u - y_1)_+^{-\left(\frac{1}{2}+\frac{1-H}{q}\right)} \ldots (u - y_q)_+^{-\left(\frac{1}{2}+\frac{1-H}{q}\right)} du \right) \right.$$

$$\left. dB(y_1) \ldots dB(y_q), t \geq 0 \right).$$

Thus, for every $c > 0$,

$$\left(Z_{ct}^{H,q}, t \geq 0 \right) \equiv^{(d)} \left(c^H Z_t^{H,q}, t \geq 0 \right).$$

To prove the stationarity of the increments, we write, for $h > 0$

$$Z_{t+h}^{H,q} - Z_t^{H,q}$$

$$= c(H, q) \int_{\mathbb{R}^q} \left(\int_t^{t+h} (u - y_1)_+^{-\left(\frac{1}{2}+\frac{1-H}{q}\right)} \ldots (u - y_q)_+^{-\left(\frac{1}{2}+\frac{1-H}{q}\right)} du \right)$$

$$\times dB(y_1) \ldots dB(y_q)$$

$$= c(H, q) \int_{\mathbb{R}^q} \left(\int_0^t (u - (y_1 - h))_+^{-\left(\frac{1}{2}+\frac{1-H}{q}\right)} \ldots (u - (y_q - h))_+^{-\left(\frac{1}{2}+\frac{1-H}{q}\right)} du \right)$$

$$\times dB(y_1) \ldots dB(y_q)$$

$$= c(H, q) \int_{\mathbb{R}^q} \left(\int_0^t (u - y_1)_+^{-\left(\frac{1}{2}+\frac{1-H}{q}\right)} \ldots (u - y_q)_+^{-\left(\frac{1}{2}+\frac{1-H}{q}\right)} du \right)$$

$$\times dB(y_1 + h) \ldots dB(y_q + h)$$

$$\equiv^{(d)} c(H, q) \int_{\mathbb{R}^q} \left(\int_0^t (u - y_1)_+^{-\left(\frac{1}{2}+\frac{1-H}{q}\right)} \ldots (u - y_q)_+^{-\left(\frac{1}{2}+\frac{1-H}{q}\right)} du \right)$$

$$\times dB(y_1) \ldots dB(y_q).$$

The last relation follows from the definition of the multiple stochastic integrals and from the fact that $(B(y + h) - B(h), y \in \mathbb{R})$ has the same finite dimensional distributions as $(B(y), y \in \mathbb{R})$. ∎

2.2.2 Moments and Hölder Continuity

The scaling property of the Hermite process implies that for every $t \geq 0$, and for every $p \geq 1$,

$$\mathbf{E}|Z_t^{H,q}|^p = t^{Hp}\mathbf{E}|Z_1^{H,q}|^p$$

and by using the stationarity of the increments, for every $0 \leq s \leq t$, and $p \geq 1$,

$$\mathbf{E}\left|Z_t^{H,q} - Z_s^{H,q}\right|^p = \mathbf{E}\left|Z_{t-s}^{H,q}\right|^p = |t-s|^{Hp}\mathbf{E}|Z_1^{H,q}|^p. \tag{2.8}$$

Via the Kolmogorov continuity criterion, we get the existence of a modification of $Z^{H,q}$, which has Hölder continuous paths of order δ, for any $\delta \in (0, H)$.

2.2.3 The Hermite Noise and the Long Memory

Consider the sequence $(X_j, j \geq 0)$ given by

$$X_j = Z_{j+1}^{H,q} - Z_j^{H,q}, \quad j \in \mathbb{N}. \tag{2.9}$$

Then, for any $j, k \in \mathbb{N}$,

$$\begin{aligned}
\mathbf{E}X_j X_k &= \mathbf{E}\left(Z_{j+1}^{H,q} - Z_j^{H,q}\right)\left(Z_{k+1}^{H,q} - Z_k^{H,q}\right) \\
&= \frac{1}{2}\left(|j-k+1|^{2H} + |j-k-1|^{2H} - 2|j-k|^{2H}\right) =: r(j-k),
\end{aligned}$$

where

$$r(j) = \frac{1}{2}\left(|j+1|^{2H} + |j-1|^{2H} - 2|j|^{2H}\right), \quad j \in \mathbb{Z}.$$

For $|j|$ large, $r(j)$ behaves as $H(2H-1)|j|^{2H-2}$. In particular, since $H \in \left(\frac{1}{2}, 1\right)$,

$$\sum_{j \in \mathbb{Z}} |r(j)| \text{ diverges.} \tag{2.10}$$

The property (2.10) is usually interpreted by saying that the Hermite process $Z^{H,q}$ has long memory (or long range dependence).

2.2.4 p-Variation

For $t > 0$, let

$$V_N^{(2)}(Z^{H,q})_t = \sum_{i=0}^{N-1} \left(Z_{t_{i+1}}^{H,q} - Z_{t_i}^{H,q} \right)^2$$

with $t_i = \frac{it}{N}$ for $i = 0, ..., N$. By (2.8),

$$\mathbf{E} V_N(Z^{H,q})_t = \mathbf{E}|Z_1^{H,q}|^2 \sum_{i=0}^{N-1} |t_{i+1} - t_i|^{2H} = t^{2H} N^{1-2H} \to_{N \to \infty} 0.$$

This implies that for every $t > 0$, the sequence $\left(V_N^{(2)}(Z^{H,q})_t, N \geq 1 \right)$ converges to 0 in $L^1(\Omega)$, i.e. the Hermite process is a zero quadratic variation process. In particular, this also implies that the Hermite process is not a semimartingale.

We can also state a result concerning the p -variation of the Hermite process. For $p \geq 1$, let

$$V_N^{(p)}(Z^{H,q})_t = \sum_{i=0}^{N-1} \left| Z_{t_{i+1}}^{H,q} - Z_{t_i}^{H,q} \right|^p \tag{2.11}$$

with $t_i, i = 0, .., N$ as before. The behavior of the sequence (2.11) is based on the fact that the Hermite noise defined by (2.9) is ergodic, and in particular it satisfies (see [37])

$$\frac{1}{N} \sum_{j=1}^{N} f(X_j) \to_{N \to \infty} \mathbf{E} f(X_1) \text{ almost surely and in } L^1(\Omega) \tag{2.12}$$

for every measurable function $f : \mathbb{R} \to \mathbb{R}$ such that $\mathbf{E}|f(X_1)| < \infty$.

Proposition 2.3 *Let $t > 0$. Consider the sequence $(V_N^{(p)}(Z^{H,q})_t, N \geq 1$ given by (2.11). Then in probability*

$$V_N^{(p)}(Z^{H,q})_t \to_{N \to \infty} \begin{cases} 0, & \text{if } p > \frac{1}{H} \\ t\mathbf{E}|Z_1^{H,q}|^{\frac{1}{H}} \text{ if } p = \frac{1}{H} \\ +\infty \text{ if } p < \frac{1}{H}. \end{cases}$$

Proof We define the sequence

$$Y_t^{N,p} = N^{pH-1} V_N^{(p)}(Z^{H,q})_t.$$

By Proposition 2.2, we have ($=^{(d)}$ means equality in distribution in the sequel),

$$Y_t^{N,p} =^{(d)} t^{Hp} \frac{1}{N} \sum_{j=0}^{N-1} |Z_{i+1}^{H,q} - Z_i^{H,q}|^p.$$

The ergodicity of the sequence (2.9) implies that (see (2.12))

$$\frac{1}{N} \sum_{j=0}^{N-1} |Z_{i+1}^{H,q} - Z_i^{H,q}|^p \rightarrow_{N\to\infty} \mathbf{E}|Z_1^{H,q}|^p \text{ almost surely and in } L^1(\Omega).$$

Consequently, $(Y_t^{N,p}, N \geq 1)$ converges in law (and thus in probability) as $N \to \infty$ to $t^{Hp}\mathbf{E}|Z_1^{H,q}|_p$. This fact easily gives the conclusion.   ∎

2.2.5 Approximation by Semimartingales

For $\varepsilon > 0$, define

$$
\begin{aligned}
Z_t^{H,q,\varepsilon} \\
= c(H,q) \int_{\mathbb{R}^q} \left(\int_0^t (u-y_1)_+^{-\left(\frac{1}{2}+\frac{1-H}{q}\right)} \ldots (u-y_q)_+^{-\left(\frac{1}{2}+\frac{1-H}{q}\right)} du \mathbf{1}_{u-y_1>\varepsilon} \ldots \mathbf{1}_{u-y_q>\varepsilon} \right) \\
dB(y_1)\ldots dB(y_q), \qquad t \geq 0.
\end{aligned}
\tag{2.13}
$$

We will show that $(Z^{H,q,\varepsilon}, \varepsilon > 0)$ constitutes a family of semi-martingales (more exactly, of finite-variation processes) and that for every $t \geq 0$, $Z^{H,q,\varepsilon}$ converges to $Z_t^{H,q}$ in $L^2(\Omega)$. We first notice that by using Fubini theorem,

$$Z_t^{H,q,\varepsilon} = \int_0^t X_{\varepsilon,u} du \tag{2.14}$$

with

$$X_{\varepsilon,u} = \int_{\mathbb{R}^q} \prod_{j=1}^q (u-y_j)_+^{-\left(\frac{1}{2}+\frac{1-H}{q}\right)} \mathbf{1}_{u-y_j>\varepsilon} dB(y_1)\ldots dB(y_q).$$

Indeed, we can write

$$X_{\varepsilon,u} = I_q(g_{\varepsilon,u})$$

with

$$g_{\varepsilon,u}(y_1, \ldots, y_q) = c(H,q) \prod_{j=1}^q (u-y_j)_+^{-\left(\frac{1}{2}+\frac{1-H}{q}\right)} \mathbf{1}_{u-y_j>\varepsilon},$$

and via the isometry property (1.16), we have for $0 \leq u \leq t$

$$\mathbf{E}X_{\varepsilon,u}^2 = q! \|g_{\varepsilon,u}\|_{L^2(\mathbb{R}^q)}^2$$

$$= q! c(H,q)^2 \int_{\mathbb{R}^q} dy_1 \ldots dy_q \left(\int_{-\infty}^{u-\varepsilon} (u-y)_+^{-2\left(\frac{1}{2}+\frac{1-H}{q}\right)} dy \right)^q$$

$$= q! c(H,q)^2 \frac{q}{2(1-H)} \varepsilon^{-\frac{2(1-H)}{q}} < \infty.$$

Consequently, by (2.14), for every $\varepsilon > 0$, the process $(Z_t^{H,q,\varepsilon}, t \geq 0)$ is a semi-martingale. We will show that it approximates the Hermite process $Z^{H,q}$ at any fixed time t.

Proposition 2.4 *For every $t \geq 0$,*

$$Z_t^{H,q,\varepsilon} \to_{\varepsilon \to 0} Z_t^{H,q} \text{ in } L^2(\Omega).$$

Proof We have

$$\mathbf{E} \left| Z_t^{H,q,\varepsilon} - Z_t^{H,q} \right|^2 = \mathbf{E} \left| Z_t^{H,q,\varepsilon} \right|^2 - 2\mathbf{E}Z_t^{H,q,\varepsilon} Z_t^{H,q} + t^{2H}.$$

By proceeding as in the proof of Proposition 2.1, we find

$$\mathbf{E} \left| Z_t^{H,q,\varepsilon} \right|^2$$

$$= q! c(H,q)^2 \int_0^t \int_0^t du dv \left(\int_{-\infty}^{(u-\varepsilon)\wedge(v-\varepsilon)} (u-y)^{-\left(\frac{1}{2}+\frac{1-H}{q}\right)} (v-y)^{-\left(\frac{1}{2}+\frac{1-H}{q}\right)} dy \right)^q$$

and

$$\mathbf{E}Z_t^{H,q,\varepsilon} Z_t^{H,q}$$

$$= q! c(H,q)^2 \int_0^t \int_0^t du dv \left(\int_{-\infty}^{(u-\varepsilon)\wedge v} (u-y)^{-\left(\frac{1}{2}+\frac{1-H}{q}\right)} (v-y)^{-\left(\frac{1}{2}+\frac{1-H}{q}\right)} dy \right)^q.$$

By the dominated convergence theorem and relation (2.3), both quantities above converge, as $\varepsilon \to 0$, to

$$q! c(H,q)^2 \int_0^t \int_0^t du dv \left(\int_{-\infty}^{u\wedge v} (u-y)^{-\left(\frac{1}{2}+\frac{1-H}{q}\right)} (v-y)^{-\left(\frac{1}{2}+\frac{1-H}{q}\right)} dy \right)^q = t^{2H}.$$

∎

2.3   Some Particular Hermite Processes: Fractional Brownian Motion and the Rosenblatt Process

The first two Hermite processes are the fractional Brownian motion and the Rosenblatt process. We focus on them in this paragraph.

2.3.1   Fractional Brownian Motion

The fractional Brownian motion is obtained by taking $q = 1$ in (2.6). That is, for $H \in \left(\frac{1}{2}, 1\right)$, the fBm $(B_t^H, t \geq 0)$ with Hurst parameter H is given by

$$B_t^H = c(H, 1) \int_{\mathbb{R}} \left(\int_0^t (u - y)_+^{H - \frac{3}{2}} du \right) dB(y), \quad t \geq 0.$$

with $c(H, 1)$ from (2.5). Since the above integral is a standard Wiener integral with respect to the Wiener process (see Sect. 1.1.1, it follows that B^H is a centered Gaussian process with covariance (2.7). It is well-known that the fBm can actually be defined for all $H \in (0, 1)$.

In this case, the covariance function (2.7) characterizes the law of the process B^H, which is not true when $q \geq 2$.

An equivalent definition of the fBm is as the only self-similar Gaussian process with stationary increments. For $H \in (0, 1)$, $H \neq \frac{1}{2}$, the fBm is neither a semimartingale, nor a Markov process.

2.3.2   The Rosenblatt Process

The Rosenblatt process is obtained by taking $q = 2$ in the relation (2.6), so

$$Z_t^{H,2} := Z_t = c(H, 2) \int_{\mathbb{R}} \int_{\mathbb{R}} \left(\int_0^t (s - y_1)_+^{-\frac{2-H}{2}} (s - y_2)_+^{-\frac{2-H}{2}} ds \right) dB(y_1) dB(y_2)$$

$$\tag{2.15}$$

where $(B(y), y \in \mathbb{R})$ is a standard Brownian motion on $\mathbb{R}$ and $c(H, 2)$ is defined in (2.5). The Rosenblatt process lives in the second Wiener chaos.

A particularity of the Rosenblatt process is that its probability law is determined by its cumulants. We recall that when $G = I_2(f)$ is a multiple integral of order 2 with respect to a Wiener sheet $(B(y), y \in \mathbb{R})$, then its cumulants can be computed via the formula (1.39) in Theorem 1.3.

Let $\lambda_1, ..., \lambda_N \in \mathbb{R}$ and $t_1, ..., t_N \geq 0$. Let

$$V = \lambda_1 Z_{t_1}^{H,2} + ... \lambda_N Z_{t_N}^{H,2}.$$

Then the cumulants of the random variable V, which characterize the finite dimensional distributions of the Rosenblatt process, can be computed as follows.

$$k_1(V) = \mathbf{E}V = 0,$$

$$k_2(V) = Var(V) = \sum_{i,j=1}^{N} \lambda_i \lambda_j \mathbf{E}\left(Z_{t_i}^{H,2} Z_{t_j}^{H,2}\right)$$

$$= \sum_{i,j=1}^{N} \lambda_i \lambda_j R^H(t_i, t_j),$$

with R^H given by (2.7). For $m \geq 3$, since

$$V = I_2\left(\sum_{j=1}^{N} \lambda_j L_{t_j}^{H,2}\right),$$

we find, by using (1.39),

$$
\begin{aligned}
&k_m(V) \\
&= 2^{m-1}(m-1)! \int_{\mathbb{R}^m} dy_1 \dots dy_m \\
&\quad \times \left(\sum_{j_1=1}^{N} L_{t_{j_1}}^{H,2}(y_1, y_2)\right)\left(\sum_{j_2=1}^{N} L_{t_{j_2}}^{H,2}(y_2, y_3)\right) \dots \left(\sum_{j_m=1}^{N} L_{t_{j_m}}^{H,2}(y_m, y_1)\right) \\
&= 2^{m-1}(m-1)! c(H,2)^m \int_{\mathbb{R}^m} dy_1 \dots dy_m \sum_{j_1,\dots,j_m=1}^{N} \lambda_{j_1} \dots \lambda_{j_m} \\
&\quad \times \left(\int_0^{t_{j_1}} (u_1 - y_1)_+^{\frac{H}{2}-1}(u_1 - y_2)_+^{\frac{H}{2}-1} du_1\right) \\
&\quad \left(\int_0^{t_{j_2}} (u_2 - y_2)_+^{\frac{H}{2}-1}(u_1 - y_3)_+^{\frac{H}{2}-1} du_2\right) \\
&\quad \times \dots \left(\int_0^{t_{j_m}} (u_m - y_m)_+^{\frac{H}{2}-1}(u_m - y_1)_+^{\frac{H}{2}-1} du_m\right).
\end{aligned}
$$

By Fubini,

$$
\begin{aligned}
k_m(V) &= 2^{m-1}(m-1)! c(H,2)^m \sum_{j_1,\dots,j_m=1}^{N} \lambda_{j_1} \dots \lambda_{j_m} \int_0^{t_{j_1}} du_1 \dots \int_0^{t_{j_m}} du_m \\
&\quad \times \prod_{a=1}^{m} \left(\int_{\mathbb{R}} (u_a - y)_+^{\frac{H}{2}-1}(u_{a+1} - y)_+^{\frac{H}{2}-1} dy\right)
\end{aligned}
$$

with the convention $u_{m+1} = u_1$. Lemma 2.2 gives

$$k_m(V)$$

$$= 2^{m-1}(m-1)!c(H,2)^m \beta\left(\frac{H}{2}, 1-H\right)^m \sum_{j_1,\ldots,j_m=1}^N \lambda_{j_1}\ldots\lambda_{j_m}$$

$$\times \int_0^{t_{j_1}} du_1 \ldots \int_0^{t_{j_m}} du_m |u_1-u_2|^{H-1}|u_2-u_3|^{H-1}\ldots|u_{m-1}-u_m|^{H-1}|u_m-u_1|^{H-1}$$

$$= 2^{\frac{m}{2}-1}(m-1)!(H(2H-1))^{\frac{m}{2}} \sum_{j_1,\ldots,j_m=1}^N \lambda_{j_1}\ldots\lambda_{j_m} \int_0^{t_{j_1}} du_1 \ldots \int_0^{t_{j_m}} du_m$$

$$\times|u_1-u_2|^{H-1}|u_2-u_3|^{H-1}\ldots|u_{m-1}-u_m|^{H-1}|u_m-u_1|^{H-1}. \tag{2.16}$$

2.4 Alternative Representation

Here, the purpose is to give a finite-interval representation of the Hermite process, i.e. to express it as a multiple Wiener-Itô integral with respect to a Wiener process index by an interval $[0, T]$ with $T > 0$. Actually, the construction of the stochastic calculus with respect to the fractional Brownian motion is based on such a finite-time interval representation (see e.g. [29]).

Let us consider the function

$$f^H(t,s) = \left(\frac{s}{t}\right)^{\frac{1}{2}-H} (t-s)_+^{H-\frac{3}{2}}, \quad s,t \geq 0,$$

and define the process $\left(Y_t^{H,q}, t \geq 0\right)$ given by

$$Y_t^{H,q} = c(H,q) \int_{[0,t]^q} \left(\int_0^t f^{H'}(u,y_1)\ldots f^{H'}(u,y_q)du\right) dW(y_1)\ldots dW(y_q) \tag{2.17}$$

where $W = (W(y), y \in \mathbb{R}_+)$ is a (standard) Wiener process, $c(H, q)$ is from (2.5) and

$$H' = 1 + \frac{H-1}{q}. \tag{2.18}$$

We can also write

$$Y_t^{H,q} = I_q(\ell_t^{H,q})$$

with

$$\ell_t^{H,q}(y_1, ..., y_q)$$

$$= c(H, q)1_{[0,t]^q}(y_1, ..., y_q) \int_0^t f^{H'}(u, y_1) \dots f^{H'}(u, y_q) du$$

$$= c(H, q)1_{[0,t]^q}(y_1, ..., y_q) \int_0^t \left(\prod_{j=1}^q \left(\frac{y_j}{u}\right)^{\frac{1}{2}-H'} (u - y_j)_+^{H'-\frac{3}{2}} \right)$$

$$= c(H, q)1_{[0,t]^q}(y_1, ..., y_q)(y_1....y_q)^{-\frac{1}{2}+\frac{1-H}{q}} \int_0^t du \ u^{\frac{q}{2}-1+H} \prod_{j=1}^q (u - y_j)^{-\frac{1}{2}+\frac{H-1}{q}}.$$

Consider the process $\left(Z_t^{H,q,\varepsilon}, t \geq 0\right)$ given by (2.13). By (2.14), we can write

$$Z_t^{H,q,\varepsilon} = c(H, q) \int_0^t du \left(\int_{\mathbb{R}^q} \prod_{j=1}^q \left((u - y_j)_+^{-\left(\frac{1}{2}+\frac{1-H}{q}\right)} 1_{u-y_j>\varepsilon}\right) dB(y_1) \dots dB(y_q) \right)$$

$$= c(H, q) \int_0^t du I_q \left((f_u^{H,\varepsilon})^{\otimes q}\right)$$

where

$$f_u^{H,\varepsilon}(y) = (u - y)_+^{-\left(\frac{1}{2}+\frac{1-H}{q}\right)} 1_{u-y>\varepsilon}.$$

Notice that for every $\varepsilon > 0$, $f_u^{H,\varepsilon} \in L^2(\mathbb{R})$, see Sect. 2.2.5. Let us also introduce the regularized fractional process

$$\dot{B}_u^{H,\varepsilon} = \int_{\mathbb{R}} (u - y)_+^{-\left(\frac{1}{2}+\frac{1-H}{q}\right)} 1_{u-y>\varepsilon} dB(y). \tag{2.19}$$

Since for every $h \in L^2(\mathbb{R})$ with $\|h\|_{L^2(\mathbb{R}} = 1$, we have by Proposition 1.6

$$I_q(h^{\otimes q}) = q! H_q(I_1(h)) \tag{2.20}$$

where H_q is the qth Hermite polynomial, we get

$$Z_t^{H,q,\varepsilon} = q! c(H, q) \int_0^t du \left(\mathbf{E}(\dot{B}_u^{H,\varepsilon})^2\right)^{\frac{q}{2}} H_q \left(\frac{\dot{B}_u^{H,\varepsilon}}{\left(\mathbf{E}(\dot{B}_u^{H,\varepsilon})^2\right)^{\frac{1}{2}}} \right). \tag{2.21}$$

Lemma 2.3 *For every $T > 0$, the process $\left(\dot{B}_u^{H,\varepsilon}, u \in [0, T]\right)$ given by (2.19) has the same finite dimensional distributions as the process $\left(X_u^{H,\varepsilon}, u \in [0, T]\right)$, where*

$$X_u^{H,\varepsilon} = \int_0^\infty z^{\frac{1}{2}-H'}(1-uz)^{H'-\frac{3}{2}}dz 1_{z<\frac{1}{\varepsilon+s}}dW(z).$$

Proof Notice that both $\dot{B}^{H,\varepsilon}$ and $X^{H,\varepsilon}$ are centered Gaussian processes, thus it suffices to check that their covariances coincide. For $0 \le v < u \le T$, we have

$$\mathbf{E}\dot{B}_u^{H,\varepsilon}\dot{B}_v^{H,\varepsilon} = \int_{\mathbb{R}} (u-y)_+^{-\left(\frac{1}{2}+\frac{1-H}{q}\right)}1_{u-y>\varepsilon}(v-y)_+^{-\left(\frac{1}{2}+\frac{1-H}{q}\right)}1_{v-y>\varepsilon}dy$$

$$= \int_{-\infty}^{v-\varepsilon}(u-y)^{-\left(\frac{1}{2}+\frac{1-H}{q}\right)}(v-y)^{-\left(\frac{1}{2}+\frac{1-H}{q}\right)}dy$$

$$= (u-v)^{2H'-2}\int_0^\infty z^{H'-\frac{3}{2}}(1+z)^{H'-\frac{3}{2}}1_{z>\frac{\varepsilon}{u-v}}dz$$

where we performed the change of variables $z = \frac{v-y}{u-v}$. With the new change of variables $z = \frac{1-x}{x}$, we get

$$\mathbf{E}\dot{B}_u^{H,\varepsilon}\dot{B}_v^{H,\varepsilon} = (u-v)^{2H'-2}\int_0^\infty x^{1-2H'}(1-x)^{H'-\frac{3}{2}}1_{x<\frac{u-v}{\varepsilon-(u-v)}}dx.$$

Thew two successive changes of variable $x = \frac{a(u-v)}{v(u-a)}$ and $a = uvb$ give

$$\mathbf{E}\dot{B}_u^{H,\varepsilon}\dot{B}_v^{H,\varepsilon} = (uv)^{H'-\frac{1}{2}}\int_0^v a^{1-2H'}(v-a)^{H'-\frac{3}{2}}(u-a)^{H'-\frac{3}{2}}1_{a<\frac{u-v}{\varepsilon+v}}$$

$$= \int_0^\infty b^{1-2H'}(1-bu)^{H'-\frac{3}{2}}(1-bv)^{H'-\frac{3}{2}}1_{b<\frac{1}{\varepsilon+u}}1_{b<\frac{1}{\varepsilon+v}}db$$

$$= \mathbf{E}X_u^{H,\varepsilon}X_v^{H,\varepsilon}.$$

∎

By (2.21) and Lemma 2.3, we can write

$$Z_t^{H,q,\varepsilon} \equiv^{(d)} q!c(H,q)\int_0^t du\,\left(\mathbf{E}(X_u^{H,\varepsilon})^2\right)^{\frac{q}{2}}H_q\left(\frac{X_u^{H,\varepsilon}}{\left(\mathbf{E}(X_u^{H,\varepsilon})^2\right)^{\frac{1}{2}}}\right).$$

Via (2.20), for $T > 0$,

$$\left(Z_t^{H,q,\varepsilon}, t \in [0,T]\right) \equiv^{(d)} \left(X_t^{H,q,\varepsilon}, t \in [0,T]\right) \tag{2.22}$$

with

$$
X_t^{H,q,\varepsilon} = c(H,q) \int_{(0,\infty)^q} \left(\int_0^t du \left(\prod_{j=1}^q z_j^{\frac{1}{2}-H'} (1 - uz_j)^{H'-\frac{3}{2}} dz 1_{z_j < \frac{1}{\varepsilon+s}} \right) \right)
$$
$$
\times dW(z_1) \dots dW(z_q).
$$

Let us state and prove a first representation of the Hermite process, different by (2.6).

Proposition 2.5 *For every $T > 0$, the Hermite process $\left(Z_t^{H,q}, t \in [0,T] \right)$ has the same finite dimensional distributions as the process $\left(X_t^{H,q}, t \in [0,T] \right)$ given by*

$$
X_t^{H,q} = c(H,q) \int_{(0,\infty)^q} \left(\int_0^t du \left(\prod_{j=1}^q z_j^{\frac{1}{2}-H'} (1 - uz_j)_+^{H'-\frac{3}{2}} dz \right) \right) dW(z_1) \dots dW(z_q).
$$

Proof Let us first notice that the process $X^{H,q}$ is well-defined. Indeed, for every $t \geq 0$,

$$
\mathbf{E}(X_t^{H,q})^2
$$
$$
= q! c(H,q)^2 \int_{(0,\infty)^q} \left(\int_0^t du \prod_{j=1}^q z_j^{\frac{1}{2}-H'} (1 - uz_j)_+^{H'-\frac{3}{2}} \right)^2 dz_1 ... dz_j
$$
$$
= q! c(H,q)^2 \lim_{\varepsilon \to 0} \int_{(0,\infty)^q} \left(\int_0^t du \prod_{j=1}^q z_j^{\frac{1}{2}-H'} (1 - uz_j)^{H'-\frac{3}{2}} 1_{z_j < \frac{1}{\varepsilon+u}} \right)^2 dz_1 ... dz_j
$$
$$
= q! c(H,q)^2 \lim_{\varepsilon \to 0} \int_{\mathbb{R}^q} \left(\int_0^t \prod_{j=1}^q (u - x_j)^{H'-\frac{3}{2}} 1_{u-x_j > \varepsilon} \right)^2 dx_1 ... dx_q
$$
$$
= q! c(H,q)^2 \int_{\mathbb{R}^q} \left(\int_0^t \prod_{j=1}^q (u - x_j)_+^{H_0-\frac{3}{2}} \right)^2 dx_1 ... dx_q.
$$

We know from Proposition 2.4 that for every $t \geq 0$,

$$
Z_t^{H,q,\varepsilon} \to_{\varepsilon \to 0} Z_t^{H,q} \text{ in } L^2(\Omega).
$$

Using the same arguments, it is easy to show that for every $t \geq 0$,

$$
X_t^{H,q,\varepsilon} \to_{\varepsilon \to 0} X_t^{H,q} \text{ in } L^2(\Omega).
$$

We conclude by taking the limit as $\varepsilon \to 0$ in (2.22). ∎

Now, set

$$W_t^{H,q,\varepsilon} = c(H,q) \int_{(0,\infty)^q} \left(\int_0^t du \left(\prod_{j=1}^q z_j^{\frac{1}{2}-H'} (1-uz_j)^{H'-\frac{3}{2}} dz 1_{z_j < \frac{1-\varepsilon}{s}} \right) \right)$$
$$\times dW(z_1)\dots dW(z_q)$$

and observe that for every $t \geq 0$,

$$W_t^{H,q,\varepsilon} \to_{\varepsilon\to 0} X_t^{H,q} \text{ in } L^2(\Omega) \tag{2.23}$$

and

$$W_t^{H,q,\varepsilon} = q! c(H,q) \int_0^t du \left(\mathbf{E}(W_u^{H,\varepsilon})^2 \right)^{\frac{q}{2}} H_q \left(\frac{W_u^{H,\varepsilon}}{\left(\mathbf{E}(W_u^{H,\varepsilon})^2 \right)^{\frac{1}{2}}} \right)$$

with

$$W_u^{H,\varepsilon} = \int_0^\infty z^{\frac{1}{2}-H'} (1-sz)^{H'-\frac{3}{2}} 1_{z < \frac{1-\varepsilon}{s}} dW(z).$$

Lemma 2.4 *The process* $(W_u^{H,\varepsilon}, u \in [0,T])$ *has the same finite dimensional distributions as* $(Y_u^{H,\varepsilon}, u \in [0,T])$, *where*

$$Y_u^{H,\varepsilon} = u^{H'-\frac{1}{2}} \int_0^\infty x^{\frac{1}{2}-H'} (u-x)^{H'-\frac{3}{2}} 1_{x < \frac{s}{1-\varepsilon}} dB(x)$$

where $(B(x), x \in [0,T])$ *is a Brownian motion.*

Proof Let $0 \leq u_1 \leq u_2 \leq T$. Then

$$\mathbf{E} W_{u_1}^{H,\varepsilon} W_{u_2}^{H,\varepsilon}$$
$$= \int_0^\infty dz z^{\frac{1}{2}-H'} (1-u_1 z)^{H'-\frac{3}{2}} 1_{z < \frac{1-\varepsilon}{u_1}} z^{\frac{1}{2}-H'} (1-u_2 z)^{H'-\frac{3}{2}} 1_{z < \frac{1-\varepsilon}{u_2}}$$
$$= (u_1 u_2)^{H'-\frac{1}{2}} \int_0^\infty x^{1-2H'} (u_1-x)^{H'-\frac{3}{2}} (u_2-x)^{H'-\frac{3}{2}} 1_{x < u_1(1-\varepsilon)} 1_{x < u_2(1-\varepsilon)}$$

where we performed the change of variables $u_1 u_2 z = x$. The last quantity coincides with $\mathbf{E} Y_{u_1}^{H,\varepsilon} Y_{u_2}^{H,\varepsilon}$. ∎

As a consequence of Lemma 2.4, $(W_t^{H,q,\varepsilon}, t \in [0,T])$ has the same finite dimensional distributions as $(Y_t^{H,q,\varepsilon}, t \in [0,T])$, where

$$Y_t^{H,q,\varepsilon} = q! c(H,q) \int_0^t du \left(\mathbf{E}(Y_u^{H,\varepsilon})^2\right)^{\frac{q}{2}} H_q \left(\frac{Y_u^{H,\varepsilon}}{\left(\mathbf{E}(Y_u^{H,\varepsilon})^2\right)^{\frac{1}{2}}} \right).$$

It can be shown that, as $\varepsilon \to 0$, for every $t \in [0, T]$, $Y_t^{H,q,\varepsilon}$ converges to $Y_t^{H,q}$. Thus, by these facts and (2.23),

$$\left(X_t^{H,q}, t \in [0, T] \right) \equiv^{(d)} \left(Y_t^{H,q}, t \in [0, T] \right)$$

where $Y^{H,q}$ is given by (2.17).

From the above identity and Proposition 2.5, we obtain the alternative representation of the Hermite process on a finite interval.

Proposition 2.6 *The Hermite process $(Z_t^{H,q}, t \in [0, T])$ has the same finite dimensional distributions as $(Y_t^{H,q}, t \in [0, T])$ given by (2.17).*

2.5   On the Simulation of the Rosenblatt Process

Let us finish this chapter with some brief comments concerning the numerical simulation of the Hermite process. Concerning the fractional Brownian motion, various techniques have been proposed in order to simulate this Gaussian stochastic process. We refer to [15] for a survey of these methods. On the other hand, the simulation of the Hermite process of order $q \geq 2$ is much more challenging and only few attempts have been done in the case $q = 2$ (i.e. for the Rosenblatt process).

- Concerning the numerical evaluation of the Rosenblatt distribution (i.e. the distribution of the random variable $Z_1^{H,2}$ given by (2.15)), the authors or [50] developed a technique to simulate its probability density function and its cumulative distribution function. This technique is based on the representation (1.36) of $Z_1^{H,2}$ as an infinite sum of shifted chi-square random variables, which holds true since $Z_1^{H,2}$ belongs to the second Wiener chaos.
- Regarding the simulation of the sample paths of the Rosenblatt process, in the work [1] the authors proposed a method based on some wavelet-type approximation of the Rosenblatt process (see also [4, 31] for related theoretical aspects concerning this approximation). The authors of [44] proposed an algorithm which uses a Donsker-type approximation of the Rosenblatt process by some perturbed random walks.
- When $q \geq 3$, although there not yet a concrete simulation scheme for the Hermite process $Z^{(q,H)}$, some theoretical concerning its wavelet expansion (with potential application to numerical analysis) have been done recently in [4, 5].

Chapter 3
The Wiener Integral with Respect to the Hermite Process and the Hermite Ornstein-Uhlenbeck Process

The stochastic integration theory with respect to the Hermite processes is at its beginning. When the integrand considered is deterministic, i.e. $f : \mathbb{R} \to \mathbb{R}$ and belongs to a suitable class of functions, the construction of its integral with respect to a Hermite process

$$\int_{\mathbb{R}} f(s) dZ_s^{H,q} \tag{3.1}$$

is pretty natural and it will be detailed below. For stochastic integrands, the theory is well developed with respect to the fractional Brownian motion (see, e.g. [29]). There are also few works concerning the stochastic integration with respect to the Rosenblatt process process (see [13, 47]).

This chapter is organized in the following way:

- We start with the constructed of the Wiener integral with respect to the Hermite process of order $q \geq 1$. This stochastic integral is defined as an isometry between a suitable space of deterministic functions and $L^2(\Omega)$.
- Then, we discuss the particular cases $q = 1$ and $q = 2$, when the integrator is the fractional Brownian motion or the Rosenblatt process. In theses situations, we characterize the law of the Wiener integral.
- We also define the Wiener integral with respect to the Hermite process in the Riemann-Stieljes sense. This definition is based on the properties of the sample paths of the Hermite process. This integral coincides with the isometric Wiener integral on the intersection of their domains.
- As an application, we define and analyze the Ornstein-Uhlenbeck process associated to the Hermite process. This stochastic process solves a Langevin-type equation with Hermite noise and it can be written as a Wiener integral with respect to the Hermite process.

© The Author(s), under exclusive license to Springer Nature Switzerland AG 2023
C. Tudor, *Non-Gaussian Selfsimilar Stochastic Processes*,
SpringerBriefs in Probability and Mathematical Statistics,
https://doi.org/10.1007/978-3-031-33772-7_3

3.1　Wiener Integral

We start by introducing the spaces of Wiener integrands. For $H \in \left(\frac{1}{2}, 1\right)$, we denote by $|\mathcal{H}_H|$ the set of measurable functions $f : \mathbb{R} \to \mathbb{R}$ such that

$$\|f\|^2_{|\mathcal{H}_H|} := \int_{\mathbb{R}} \int_{\mathbb{R}} |f(u)| \cdot |f(v)||u - v|^{2H-2} du dv < \infty.$$

The set $|\mathcal{H}_H|$ is a Banach space with respect to the norm $\| \cdot \|_{|\mathcal{H}_H|}$, see e.g. [32].

Let us also introduce the space $\mathcal{H}_H$ of measurable functions $f : \mathbb{R} \to \mathbb{R}$ such that $\|f\|_{\mathcal{H}_H} < \infty$, where we set

$$\langle f, g \rangle_{\mathcal{H}_H} = H(2H - 1) \int_{\mathbb{R}} \int_{\mathbb{R}} f(u)g(v)|u - v|^{2H-2} du dv$$

and

$$\|f\|^2_{\mathcal{H}_H} = H(2H - 1) \int_{\mathbb{R}} \int_{\mathbb{R}} f(u)f(v)|u - v|^{2H-2} du dv. \tag{3.2}$$

We have the inclusion

$$|\mathcal{H}_H| \subset \mathcal{H}_H.$$

In particular, the indicator function $1_{[0,t]}$ belongs to $|\mathcal{H}_H|$ for every $t > 0$ and for $s, t \geq 0$,

$$\langle 1_{[0,t]}, 1_{[0,s]} \rangle_{|\mathcal{H}_H|} = H(2H - 1) \int_0^t \int_0^s |u - v|^{2H-2} dv du = R^H(t, s), \tag{3.3}$$

due to Lemma 2.1.

The purpose is to define a stochastic integral of Wiener type, for a deterministic function f in a suitable class of functions, with respect to the Hermite process. First, let us take f to be step function of the form

$$f(t) = \sum_{i=0}^{N-1} \lambda_i 1_{[t_i, t_{i+1})}(t) \tag{3.4}$$

with $\lambda_i \in \mathbb{R}$ for $i = 0, ..., N - 1$ and with $0 = t_0 < t_1 < ... < t_N$. For f as in (3.4), we set

$$\int_{\mathbb{R}} f(u) dZ_u^{H,q} = \sum_{i=0}^{N-1} \lambda_i \left(Z_{t_{i+1}}^{H,q} - Z_{t_i}^{H,q} \right).$$

Let us calculate the $L^2(\Omega)$-norm of the above Wiener integral. We have, by using (2.7),

$$\mathbf{E}\left(\int_{\mathbb{R}} f(u)dZ_u^{H,q}\right)^2$$

$$= \mathbf{E}\left(\sum_{i=0}^{N-1} \lambda_i \left(Z_{t_{i+1}}^{H,q} - Z_{t_i}^{H,q}\right)\right)^2$$

$$= \sum_{i,j=0}^{N-1} \lambda_i\lambda_j \mathbf{E}\left(Z_{t_{i+1}}^{H,q} - Z_{t_i}^{H,q}\right)\left(Z_{t_{j+1}}^{H,q} - Z_{t_j}^{H,q}\right)$$

$$= \sum_{i,j=0}^{N-1} \lambda_i\lambda_j \left[R^H(t_{i+1}, t_{j+1}) - R^H(t_{i+1}, t_j) - R^H(t_i, t_{j+1}) + R^H(t_i, t_j)\right].$$

On the other hand, by (3.3),

$$\|f\|_{\mathcal{H}_H}^2 = \sum_{i,j=0}^{N-1} \lambda_i\lambda_j \langle 1_{[t_i,t_{i+1})}, 1_{[t_j,t_{j+1})}\rangle_{\mathcal{H}_H}$$

$$= \sum_{i,j=0}^{N-1} \lambda_i\lambda_j \left[R^H(t_{i+1}, t_{j+1}) - R^H(t_{i+1}, t_j) - R^H(t_i, t_{j+1}) + R^H(t_i, t_j)\right]$$

$$= \mathbf{E}\left(\int_{\mathbb{R}} f(u)dZ_u^{H,q}\right)^2.$$

In order to obtain a more explicit expression of the Wiener integral, we notice that

$$\int_{\mathbb{R}} f(u)dZ_u^{H,q}$$

$$= \sum_{i=0}^{N-1} \lambda_i \left(Z_{t_{i+1}}^{H,q} - Z_{t_i}^{H,q}\right)$$

$$= c(H,q)\sum_{i=0}^{N-1} \lambda_i \int_{\mathbb{R}^q}\left[\int_{t_i}^{t_{i+1}} \prod_{j=1}^{q}(u-y_j)_+^{-\left(\frac{1}{2}+\frac{1-H}{q}\right)} du\right] dB(y_1)\dots dB(y_q)$$

$$= c(H,q)\int_{\mathbb{R}^q}\left[\int_{\mathbb{R}} du \left(\sum_{i=0}^{N-1} \lambda_i 1_{[t_i,t_{i+1})}(u)\right)\prod_{j=1}^{q}(u-y_j)_+^{-\left(\frac{1}{2}+\frac{1-H}{q}\right)}\right] dB(y_1)\dots dB(y_q)$$

$$= c(H,q)\int_{\mathbb{R}^q}\left[\int_{\mathbb{R}} du f(u)\prod_{j=1}^{q}(u-y_j)_+^{-\left(\frac{1}{2}+\frac{1-H}{q}\right)}\right] dB(y_1)\dots dB(y_q)$$

$$= I_q(Jf),$$

where

$$(Jf)(y_1, \dots y_q) = c(H, q) \int_{\mathbb{R}} du f(u) \prod_{j=1}^{q} (u - y_j)_+^{-\left(\frac{1}{2} + \frac{1-H}{q}\right)}. \qquad (3.5)$$

This motivates the following definition of the Wiener integral with respect to the Hermite process.

Definition 3.1 For a measurable function $f : \mathbb{R} \to \mathbb{R}$, we define

$$\int_{\mathbb{R}} f(u) dZ_u^{H,q} = I_q(Jf)$$

where Jf is given by (3.5), provided that $Jf \in L^2(\mathbb{R}^q)$.

We will show that the space $\mathcal{H}_H$ defined above coincides with the set of measurable functions $f : \mathbb{R} \to \mathbb{R}$ such that $Jf \in L^2(\mathbb{R}^q)$. Indeed,

$$\|Jf\|_{L^2(\mathbb{R}^q)}^2 = c(H, q)^2 \int_{\mathbb{R}^q} dy_1 \dots dy_q$$

$$\int_{\mathbb{R}} \int_{\mathbb{R}} dudv f(u) f(v) \prod_{j=1}^{q} (u - y_j)_+^{-\left(\frac{1}{2} + \frac{1-H}{q}\right)} \prod_{j=1}^{q} (v - y_j)_+^{-\left(\frac{1}{2} + \frac{1-H}{q}\right)}$$

and by Fubini,

$$\|Jf\|_{L^2(\mathbb{R}^q)}^2$$

$$= c(H, q)^2 \int_{\mathbb{R}} \int_{\mathbb{R}} dudv f(u) f(v) \left(\int_{\mathbb{R}} dy (u - y)_+^{-\left(\frac{1}{2} + \frac{1-H}{q}\right)} (v - y)_+^{-\left(\frac{1}{2} + \frac{1-H}{q}\right)} \right)^q$$

$$= c(H, q)^2 \beta \left(\frac{H}{2}, 1 - H \right)^q \int_{\mathbb{R}} \int_{\mathbb{R}} dudv f(u) f(v) |u - v|^{2H-2}$$

$$= \frac{1}{q!} H(2H - 1) \int_{\mathbb{R}} \int_{\mathbb{R}} dudv f(u) f(v) |u - v|^{2H-2},$$

where we used relation (2.3) in Lemma 2.2. Consequently, the Wiener integral $\int_{\mathbb{R}} f(u) dZ_u^{H,q}$ is well-defined for every $f \in \mathcal{H}_H$ and in particular, for any $f \in |\mathcal{H}_H|$.

Proposition 3.1 *The Wiener integral is an isometry, i.e. for every* $f, g \in |\mathcal{H}_H|$,

$$\mathbf{E} \left(\int_{\mathbb{R}} f(u) dZ_u^{H,q} \right) \left(\int_{\mathbb{R}} g(u) dZ_u^{H,q} \right) = \langle f, g \rangle_{\mathcal{H}_H}. \qquad (3.6)$$

Proof As above, we have by (1.16),

$$
\mathbf{E}\left(\int_{\mathbb{R}} f(u)dZ_u^{H,q}\right)\left(\int_{\mathbb{R}} g(u)dZ_u^{H,q}\right)
$$

$$
= q!\langle Jf, Jq\rangle_{L^2(\mathbb{R}^q)} = H(2H-1)\int_{\mathbb{R}}\int_{\mathbb{R}} dudvg(u)f(v)|u-v|^{2H-2} = \langle f, g\rangle_{\mathcal{H}_H}.
$$

∎

3.2   The Cases $q = 1$ and $q = 2$

As for the Hermite process itself, we have information on the probability distribution of the object $\int_{\mathbb{R}} f(u)dZ_u^{H,q}$, with $f \in \mathcal{H}_H$ if $q = 1$ or $q = 2$. When $q = 1$, this object constitutes the Wiener integral with respect to the fractional Brownian motion, which is Gaussian. Actually,

$$
\int_{\mathbb{R}} f(u)dZ_u^{H,q} \sim N\left(0, \|f\|_{\mathcal{H}_H}^2\right),
$$

with $\|\cdot\|_{\mathcal{H}_H}$ given by (3.2).

Let us discuss the case $q = 2$. Let $(Z_t^{H,2}, t \geq 0)$ be the Rosenblatt process and $f \in \mathcal{H}_H$. By definition, the Wiener integral $\int_{\mathbb{R}} f(u)dZ_u^{H,2}$ (which will be called as Rosenblatt-Wiener integral) is well-defined and it belongs to the second Wiener chaos. Consequently, its probability distribution will be entirely determined by its cumulants, see Proposition 1.9. We have

$$
k_1\left(\int_{\mathbb{R}} f(u)dZ_u^{H,2}\right) = \mathbf{E}\left(\int_{\mathbb{R}} f(u)dZ_u^{H,2}\right) = 0
$$

and by (3.6),

$$
k_2\left(\int_{\mathbb{R}} f(u)dZ_u^{H,2}\right) = Var\left(\int_{\mathbb{R}} f(u)dZ_u^{H,2}\right) = \|f\|_{\mathcal{H}_H}^2.
$$

For $m \geq 3$, we have by (1.39), with Jf given by (3.5),

$$
k_m\left(\int_{\mathbb{R}} f(u)dZ_u^{H,2}\right) = k_m(I_2(Jf))
$$

$$
= 2^{m-1}(m-1)!\int_{\mathbb{R}^m} dy_1...dy_m (Jf)(y_1, y_2)(Jf)(y_2, y_3)\ldots(Jf)(y_m, y_1)
$$

$$
= 2^{m-1}(m-1)!c(H, q)^m\int_{\mathbb{R}^m} dy_1...dy_m
$$

$$\times \left(\int_{\mathbb{R}} f(u_1)(u_1 - y_1)_+^{-\left(\frac{1}{2} + \frac{1-H}{q}\right)}(u_1 - y_2)_+^{-\left(\frac{1}{2} + \frac{1-H}{q}\right)} du_1 \right)$$

$$\ldots \left(\int_{\mathbb{R}} f(u_m)(u_m - y_m)_+^{-\left(\frac{1}{2} + \frac{1-H}{q}\right)}(u_m - y_1)_+^{-\left(\frac{1}{2} + \frac{1-H}{q}\right)} du_m \right)$$

$$= 2^{m-1}(m-1)!c(H,q)^m \int_{\mathbb{R}^m} du_1 \ldots du_m$$

$$\prod_{j=1}^{m} \int_{\mathbb{R}} dy_j (u_j - y_j)^{-\left(\frac{1}{2} + \frac{1-H}{q}\right)}(u_{j+1} - y_j)^{-\left(\frac{1}{2} + \frac{1-H}{q}\right)}$$

with $u_{m+1} = u_1$. By (2.3),

$$k_m \left(\int_{\mathbb{R}} f(u) dZ_u^{H,2} \right)$$

$$= 2^{m-1}(m-1)!c(H,q)^m \beta \left(\frac{H}{2} 1 - H \right)^m \int_{\mathbb{R}^m} du_1 \ldots du_m f(u_1) \ldots f(u_m)$$

$$|u_1 - u_2|^{H-1}|u_2 - u_3|^{H-1} \ldots |u_{m-1} - u_m|^{H-1}|u_m - u_1|^{H-1}$$

$$= 2^{\frac{m}{2}-1}(m-1)!(H(2H-1))^{\frac{m}{2}} \int_{\mathbb{R}^m} du_1 \ldots du_m f(u_1) \ldots f(u_m)$$

$$|u_1 - u_2|^{H-1}|u_2 - u_3|^{H-1} \ldots |u_{m-1} - u_m|^{H-1}|u_m - u_1|^{H-1}. \qquad (3.7)$$

If $f = 1_{[0,1]}$, we retrieve the formula (2.16).

Let us present an easy example (which will also appear in the sequel) of a Hermite-Wiener integral.

Example 3.1 For every $t > 0$, $\lambda \in \mathbb{R}$, the function $g(u) = 1_{[0,t]}(u)e^{\lambda u}$ belongs to $|\mathcal{H}_H|$. Indeed, for $\lambda \leq 0$,

$$\|g\|_{|\mathcal{H}_H|}^2 = \int_0^t \int_0^t du\,dv\, e^{\lambda u} e^{\lambda v} |u - v|^{2H-2}$$

$$\leq \int_0^t \int_0^t du\,dv |u - v|^{2H-2} = (H(2H-1))^{-1}t^{2H} < \infty$$

and for $\lambda > 0$,

$$\|g\|_{|\mathcal{H}_H|}^2 = \int_0^t \int_0^t du\,dv\, e^{\lambda u} e^{\lambda v} |u - v|^{2H-2}$$

$$\leq e^{2\lambda t}(H(2H-1))^{-1}t^{2H} < \infty.$$

Consequently, the Hermite-Wiener integral $\int_{\mathbb{R}} g(u) dZ_u^{H,q}$ is well-defined. We will see that it is also well-defined in the pathwise sense.

3.3   Wiener Integral in the Riemann-Stieltjes Sense

We can define the integral of a deterministic function f with respect to the Hermite process in the Riemann-Stieltjes sense, by using the regularity of the sample paths of the integrator. Let $a, b \in \mathbb{R}$ such that $-\infty < a < b < \infty$. Let $f : [a, b] \to \mathbb{R}$ be a function with bounded variation. Then the Riemann-Stieljes integral of f with respect to $Z^{H,q}$ is well-defined (it will be denoted in the sequel by $\int_a^b f(u) d_{RS} Z_u^{H,q}$) and it is given by (see Sect. 2.3 in [52])

$$\int_a^b f(u) d_{RS} Z_u^{H,q} = f(b) Z_b^{H,q} - f(a) Z_a^{H,q} - \int_a^b Z_u^{H,q} df(u),$$

where the integral $df(u)$ stands for the integral with respect to the bounded variation function f, see Chap. 6 in [36].

The construction of such a Riemann-Stieltjes integral with respect to the Hermite process can be also done on intervals of the form $(-\infty, b]$. Let $f : (-\infty, b]$ be a function of bounded variation (i.e. f is with bounded variation on any interval $[a', b'] \subset (-\infty, b] \to \mathbb{R})$ and assume that

$$\lim_{a \to -\infty} \left(f(a) Z_a^{H,q} + \int_a^b Z_u^{H,q} df(u) \right) := \tilde{L}_a \in \mathbb{R}. \tag{3.8}$$

Set

$$\int_{-\infty}^b f(u) d_{RS} Z_u^{H,q} = \lim_{a \to -\infty} \int_a^b f(u) d_{RS} Z_u^{H,q}. \tag{3.9}$$

Then (see [52], Sect. 2.3) $\int_{-\infty}^b f(u) d_{RS} Z_u^{H,q}$ is well-defined and

$$\int_{-\infty}^b f(u) d_{RS} Z_u^{H,q} = f(b) Z_b^{H,q} - \tilde{L}_a \tag{3.10}$$

In particular, if $f : [a, b] \to \mathbb{R}$ $(-\infty \leq a < b < \infty)$ is continuosly differentiable we have

$$\int_a^b f(u) d_{RS} Z_u^{H,q} = f(b) Z_b^{H,q} - L_a - \int_a^b f'(u) Z_u^{H,q} du$$

with $L_a = \lim_{u \to a} f(u) Z_u^{H,q}$.

The Wiener and Riemann-Stieljes integrals with respect to the Hermite process coincide on the intersection on their domains.

Proposition 3.2 Let $f \in |\mathcal{H}_H|$ such that $\int_a^b f(u) d_{RS} Z_u^{H,q}$ is well-defined. Then

$$\int_a^b f(u) d Z_u^{H,q} = \int_a^b f(u) d_{RS} Z_u^{H,q}. \tag{3.11}$$

Proof If f is a step function of the form (3.4) (defined via a partition of $[a, b]$), then both sides of (3.11) coincide (due to construction of the Hermite Wiener integral in Sect. 3.1 and by the definition of the pathwise integral). Then we use the fact that the step functions are dense in $|\mathcal{H}_H|$, so for any $f \in |\mathcal{H}_H|$, there exists a sequence $(f_n, n \geq 1)$ of step functions such that

$$\| f_n - f \|_{|\mathcal{H}_H|} \to_{n \to \infty} 0. \tag{3.12}$$

By the isometry of the Hermite-Wiener integral (Proposition 3.1),

$$\mathbf{E} \left(\int_a^b (f_n(u) - f(u)) dZ_u^{H,q} \right)^2$$

$$= H(2H - 1) \int_a^b \int_a^b du\,dv\, (f_n(u) - f(u))\,(f_n(v) - f(v))\,|u - v|^{2H-2}$$

$$\leq H(2H - 1) \int_a^b \int_a^b du\,dv\,|f_n(u) - f(u)|\,|f_n(v) - f(v)|\,|u - v|^{2H-2}$$

$$= \| f_n - f \|_{|\mathcal{H}_H|}^2 \to_{n \to \infty} 0.$$

On the other hand, (3.12) also implies that f_n converges to f almost everywhere when $n \to \infty$ and this implies that $\int_a^b f_n(u) d_{RS} Z_u^{H,q}$ converges almost surely to $\int_a^b f(u) d_{RS} Z_u^{H,q}$. ∎

3.4   The Hermite Ornstein-Uhlenbeck Process

3.4.1   Definition and Properties

An interesting example of a stochastic process which is defined via a Hermite-Wiener integral is the Ornstein-Uhlenbeck process with respect to the Hermite process, called in the sequel as Hermite Ornstein-Uhlenbeck process. Consider the Langevin equation

$$X_t = \xi - \lambda \int_0^t X_s ds + \sigma Z_t^{H,q}, \quad t \geq 0 \tag{3.13}$$

with $\sigma, \lambda > 0$ and with initial value $\xi \in L^2(\Omega)$. The random noise $Z^{H,q}$ is a Hermite process of order $q \geq 1$ with self-similarity index $H \in \left(\frac{1}{2}, 1\right)$.

The unique solution to (3.13) can be written as

$$X_t = e^{-\lambda t} \left(\xi + \sigma \int_0^t e^{\lambda u} dZ_u^{H,q} \right), \quad t \geq 0. \tag{3.14}$$

By Example 3.1, the Wiener integral $\int_0^t e^{\lambda u} dZ_u^{H,q}$ is well-defined in $L^2(\Omega)$. Using the expression of the Hermite Wiener integral, we can write, for every $t \geq 0$,

$$X_t = \xi e^{-\lambda t} + \sigma c(H, q) \int_{\mathbb{R}^q} dB(y_1)....dB(y_q)$$

$$\times \int_0^t du \times e^{-\lambda(t-u)} \prod_{j=1}^{q} (u - y_j)_+^{-\left(\frac{1}{2} + \frac{1-H}{q}\right)}$$

$$= e^{-\lambda t}\xi + I_q(h_t)$$

with $c(H, q)$ given by (2.5) and with

$$h_t(y_1, ..., y_q) = \sigma c(H, q) \int_0^t du \times e^{-\lambda(t-u)} \prod_{j=1}^{q} (u - y_j)_+^{-\left(\frac{1}{2} + \frac{1-H}{q}\right)}.$$

When ξ is constant, then the covariance of process X is given by, if $s, t \geq 0$,

$$Cov(X_t X_s) = \sigma^2 e^{-\lambda(t+s)} \mathbf{E}\left(\int_0^t e^{\lambda u} dZ_u^{H,q}\right) \mathbf{E}\left(\int_0^s e^{\lambda u} dZ_u^{H,q}\right)$$

$$= \sigma^2 \alpha_H e^{-\lambda(t+s)} \int_0^t \int_0^s e^{\lambda u} e^{\lambda v} |u - v|^{2H-2} du dv. \tag{3.15}$$

Proposition 3.3 *For every $T > 0$ and for every $p \geq 2$, we have*

$$\sup_{t \in [0,T]} \mathbf{E}|X_t|^p \leq C_{T,p} < \infty. \tag{3.16}$$

Proof Assume $\xi = 0$ for simplicity. By (3.15), we have

$$\mathbf{E}X_t^2 = \sigma^2 \alpha_H \int_0^t \int_0^t du dv e^{-\lambda(t-u)} e^{-\lambda(t-v)} |u - v|^{2H-2}$$

and thus, for every $T > 0$

$$\sup_{t \in [0,T]} \mathbf{E}X_t^2 \leq \sigma^2 \alpha_H \int_0^T \int_0^T du dv |u - v|^{2H-2} = \sigma^2 \alpha_H T^{2H}.$$

This bound and the hypercontractivity property (the bound (1.31) in Proposition 1.7) imply (3.16). ∎

Let $\xi \in \mathbb{R}$. When $q = 1$, then X is a Gaussian process and the covariance determines the finite dimensional distributions. When $q = 2$, then X belongs to the second Wiener chaos and its finite dimensional distributions are given by the cumulants. By using the relation (3.7), we get, since $k_m(X + c) = k_m(X)$ for $X \in L^m(\Omega)$ and $c \in \mathbb{R}$,

$$k_m(X_t) = k_m \left(\sigma \int_0^t e^{-\lambda(t-u)} dZ_u^{H,q} \right)$$

$$= 2^{m-1}(m-1)!c(H,q)^m \beta \left(\frac{H}{2}, 1-H \right)^m e^{-m\lambda t} \int_{\mathbb{R}^m} du_1 ... du_m$$

$$\times e^{\lambda(u_1+...+u_m)} |u_1 - u_2|^{H-1} .... |u_{m-1} - u_m|^{H-1} |u_m - u_1|^{H-1}.$$

Due to the Riemann-Stieltjes interpretation, it is possible to give another expression of the Hermite Ornstein-Uhlenbeck process.

Proposition 3.4 *Let* $(X_t, t \geq 0)$ *be given by (3.14). Then for every* $t \geq 0$,

$$X_t = e^{-\lambda t} \xi + \sigma Z_t^{H,q} - \sigma \lambda \int_0^t Z_u^{H,q} e^{-\lambda(t-u)} du. \tag{3.17}$$

Proof By Proposition 3.2,

$$\int_0^t e^{\lambda u} dZ_u^{H,q} = \int_0^t e^{\lambda u} d_{RS} Z_u^{H,q}.$$

Thus, due to (3.14),

$$X_t = e^{-\lambda t} \xi + \sigma e^{-\lambda t} \left(Z_t^{H,q} e^{\lambda t} - \sigma \lambda \int_{)}^t Z_u^{H,q} du \right)$$

$$= e^{-\lambda t} \xi + \sigma Z_t^{H,q} - \sigma \lambda \int_0^t Z_u^{H,q} e^{-\lambda(t-u)} du.$$

∎

Let us state and prove other properties of the Hermite Ornstein-Uhlenbeck process. They have applications to mathematical finance (see e.g. [20]).

Proposition 3.5 *Let* $(X_t, t \geq 0)$ *be given by (3.14). Then, for every* $t > 0$,

$$\mathbf{E} \left(X_t - X_0 - \sigma Z_t^{H,q} \right)^2 \to_{\lambda \to 0} 0.$$

Proof We have for $t > 0$,

$$X_t - X_0 = \xi(e^{-\lambda t} - 1) + \sigma e^{-\lambda t} \int_0^t e^{\lambda u} dZ_u^{H,q}$$

so

$$X_t - X_0 - \sigma Z_t^{H,q} = \xi(e^{-\lambda t} - 1) + \int_0^t \left(e^{-\lambda(t-u)} - 1 \right) dZ_u^{H,q}.$$

Consequently,

$$
\mathbf{E}\left(X_t - X_0 - \sigma Z_t^{H,q}\right)^2
$$

$$
= (\mathbf{E}\xi^2)(e^{-\lambda t} - 1)^2 + \mathbf{E}\left(\int_0^t \left(e^{-\lambda(t-u)} - 1\right) dZ_u^{H,q}\right)^2
$$

$$
= (\mathbf{E}\xi^2)(e^{-\lambda t} - 1)^2
$$

$$
+ H(2H - 1)\int_0^t\int_0^t \left(e^{-\lambda(t-u)} - 1\right)\left(e^{-\lambda(t-v)} - 1\right)|u - v|^{2H-2}dudv.
$$

For every u, we have that $\left(e^{-\lambda(t-u)} - 1\right)\left(e^{-\lambda(t-v)} - 1\right)|u - v|^{2H-2}1_{[0,t]}(u)1_{[0,t]}(v)$ converges to zero as $\lambda \to 0$. It suffices to apply the dominated convergence theorem. ∎

Proposition 3.6 *For every* $T > 0$, $p \geq 2$,

$$
\sup_{t\in[0,T]} \mathbf{E}\left|X_t - X_0 - \sigma Z_t^{H,q}\right|^p \to_{\lambda\to 0} 0.
$$

Proof We have by (3.17),

$$
X_t - X_0 = (e^{-\lambda t} - 1)\xi + \sigma Z_t^{H,q} - \sigma\lambda\int_0^t Z_u^{H,q}e^{-\lambda(t-u)}du
$$

and consequently,

$$
\mathbf{E}\left|X_t - X_0 - \sigma Z_t^{H,q}\right|^p
$$

$$
\leq C_p\mathbf{E}|e^{-\lambda t} - 1|^p + C_p\lambda\sigma\mathbf{E}\left|\int_0^t Z_u^{H,q}e^{-\lambda(t-u)}du\right|^p
$$

$$
\leq C_p\mathbf{E}|e^{-\lambda t} - 1|^p + C_p\lambda\sigma\left(\mathbf{E}\left|\int_0^t Z_u^{H,q}e^{-\lambda(t-u)}du\right|^2\right)^{\frac{p}{2}}
$$

where we used the hypercontractivity property (1.31). Thus, it suffices to show that

$$
\sup_{t\in[0,T]} \mathbf{E}\left|\int_0^t Z_u^{H,q}e^{-\lambda(t-u)}du\right|^2 < C_T. \tag{3.18}
$$

We have, for $t \in [0, T]$,

$$\mathbf{E}\left|\int_0^t Z_u^{H,q} e^{-\lambda(t-u)} du\right|^2 = \int_0^t \int_0^t du\,dv\, e^{-\lambda(t-u)} e^{-\lambda(t-v)} \mathbf{E} Z_u^{H,q} Z_v^{H,q}$$

$$= \int_0^t \int_0^t du\,dv\, e^{-\lambda(t-u)} e^{-\lambda(t-v)} R^H(u,v)$$

$$= \int_0^t \int_0^t du\,dv\, e^{-\lambda u} e^{-\lambda v} R^H(t-u, t-v)$$

$$\leq T^{2H} \int_0^T \int_0^T e^{-\lambda u} e^{-\lambda v} \leq T^{2H+2} = C_T$$

and implies (3.18).   ∎

3.4.2 The Stationary Hermite Ornstein-Uhlenbeck Process

Let us take the initial value

$$\xi = \sigma \int_{-\infty}^0 e^{\lambda u} dZ_u^{H,q}$$

in (3.14) and denote, for $t \geq 0$,

$$X_{0,t} = \sigma e^{-\lambda t} \int_{-\infty}^t e^{\lambda u} dZ_u^{H,q}. \tag{3.19}$$

To see that this process is well-defined, we notice that, for $t > 0$,

$$\mathbf{E} X_{0,t}^2 = \sigma^2 H(2H-1) \int_{-\infty}^t \int_{-\infty}^t du\,dv\, e^{-\lambda(t-u)} e^{-\lambda(t-v)} |u-v|^{2H-2}$$

$$= \sigma^2 H(2H-1) \int_0^\infty \int_0^\infty e^{-\lambda u} e^{-\lambda v} |u-v|^{2H-2}$$

$$= 2\sigma^2 H(2H-1) \int_0^\infty du\, e^{-\lambda u} \int_0^u dv\, e^{\lambda v} |u-v|^{2H-2}$$

$$= 2\sigma^2 H(2H-1) \int_0^\infty du\, e^{-\lambda u} u^{2H-1} \int_0^1 dz(1-z)^{2H-2} e^{-\lambda u z}$$

$$\leq C \int_0^\infty u^{2H-1} e^{-\lambda u} \leq C.$$

We show that the law of the process (3.19) is stationary. Recall that a stochastic process $(Y_t, t \geq 0)$ is stationary if and only if $(Y_{t+h}, t \geq 0)$ has the same finite-dimensional distributions as $(Y_t, t \geq 0)$.

Proposition 3.7 *The process $(X_{0,t}, t \geq 0)$ given by (3.19) is stationary.*

Proof For $h > 0$,

$$X_{0,t+h} = \sigma e^{-\lambda(t+h)} \int_{-\infty}^{t+h} e^{\lambda u} dZ_u^{H,q}$$

$$= \sigma e^{-\lambda(;t+h)} \int_{-\infty}^{t} e^{\lambda(u+h)} dZ_{u+h}^{H,q} = \sigma e^{-\lambda t} \int_{-\infty}^{t} e^{\lambda u} dZ_{u+h}^{H,q}$$

and since the Hermite process $Z^{H,q}$ has stationary increments, we notice that

$$\left(\int_{-\infty}^{t} e^{\lambda u} dZ_{u+h}^{H,q}, t \geq 0 \right) \equiv^{(d)} \left(\int_{-\infty}^{t} e^{\lambda u} dZ_u^{H,q}, t \geq 0 \right).$$

So, $(X_{0,t+h}, t \geq 0)$ and $(X_{0,t}, t \geq 0)$ have the same finite dimensional distributions. ∎

In particular, the covariance of the process $(X_{0,t}, t \geq 0)$ is given by, for $0 \leq s, t$,

$$\mathbf{E} X_{0,t} X_{0,s} = \alpha_H \sigma^2 \int_{-\infty}^{t} du \int_{-\infty}^{s} dv e^{-\lambda(t-u)} e^{-\lambda(s-v)} |u - v|^{2H-2}$$

$$= \alpha_H \sigma^2 \int_0^{\infty} \int_0^{\infty} dudv e^{-\lambda u} e^{-\lambda v} |t - s - (u - v)|^{2H-2}$$

$$= \alpha_H \sigma^2 \int_0^{\infty} \int_0^{\infty} dudv e^{-\lambda u} e^{-\lambda v} ||t - s| - (u - v)|^{2H-2}$$

so $\mathbf{E} X_{0,t} X_{0,s}$ is a function of $|t - s|$.

Chapter 4
Hermite Sheets and SPDEs

The multiparameter stochastic processes are natural mathematical objects used to modelize the evolution of a system which is not influenced only by the time, but also by other variables, such as the space location. A typical example is the so-called space-time white noise which constitutes the driving noise for many stochastic differential equations. The purpose here is to introduce a multiparameter version of the Hermite processes described in Chap. 2. We mainly follow the lines of Chaps. 2 and 3, but this time in a multidimensional context. We start by properly defining the multiparameter Hermite process (or the Hermite sheet) and then we analyze its main properties. These random fields are elements of the Wiener chaos generated by the Brownian sheet and then they can be expressed as multiple stochastic integrals with respect to this Gaussian field. We prove that the Hermite sheets also enjoy properties like self-similarity, stationarity of the increments or Hölder regularity of the trajectories, but these notions are now understood in a multidimensional manner.

We also introduce a Wiener integral with respect to the Hermite sheet, as a counterpart of the integral (3.1) described in Chap. 3. This allows to consider SPDEs driven by the an additive multiparameter Hermite processes. More precisely, one can consider SPDEs of the form

$$Lu(t, x) = \dot{Z}_{t,x}^{H,q}, \qquad t \in \mathbb{R}_+, x \in \mathbb{R}^d, \qquad (4.1)$$

where L is a first or second order operator and $\dot{Z}_{t,x}^{H,q}$ stands for the formal derivative of the Hermite sheet $Z^{H,q}$. The mild solution to (4.1) can be expressed as the Wiener integral of the Green kernel associated to the operator L with respect to the Hermite sheet $Z^{H,q}$. If one takes

$$Lu(t, x) = \frac{\partial}{\partial t} u(t, x) - \Delta u(t, x),$$

C. Tudor, *Non-Gaussian Selfsimilar Stochastic Processes*,
SpringerBriefs in Probability and Mathematical Statistics,
https://doi.org/10.1007/978-3-031-33772-7_4

(Δ denotes the standard Laplacian operator), then (4.1) is the stochastic heat equation with Hermite noise, while for

$$Lu(t, x) = \frac{\partial^2}{\partial t^2} u(t, x) - \Delta u(t, x), \tag{4.2}$$

we obtain the stochastic wave equation. In this chapter, we analyze in details the case of the heat equation driven by a Hermite sheet (other situations have been also treated in the literature, see [12] for the wave equation or [48] for the fractional heat equation). We give a necessary and sufficient condition for the existence of the solution (this condition is in terms of the Hurst parameter of the Hermite noise and of the spatial dimension d) and we study various properties of the law and of paths of this solution.

4.1 Definition of the Hermite Sheet

Since we are dealing with multi-indices stochastic process, we will introduce some notation, needed in order to facilitate the lecture. For $d \in \mathbb{N} \setminus \{0\}$ if $\mathbf{a} = (a_1, a_2, \ldots, a_d)$, $\mathbf{b} = (b_1, b_2, .., b_d)$, $\alpha = (\alpha_1, .., \alpha_d)$ are vectors in $\mathbb{R}^d$, we set

$$\mathbf{ab} = \prod_{i=1}^{d} a_i b_i, \quad |\mathbf{a} - \mathbf{b}|^\alpha = \prod_{i=1}^{d} |a_i - b_i|^{\alpha_i},$$

$$\mathbf{a}/\mathbf{b} = (a_1/b_1, a_2/b_2, \ldots, a_d/b_d), \quad [\mathbf{a}, \mathbf{b}] = \prod_{i=1}^{d} [a_i, b_i], \quad (\mathbf{a}, \mathbf{b}) = \prod_{i=1}^{d} (a_i, b_i),$$

$$\sum_{\mathbf{i}=0}^{\mathbf{N}} a_{\mathbf{i}} = \sum_{i_1=0}^{N_1} \sum_{i_2=0}^{N_2} \cdots \sum_{i_d=0}^{N_d} a_{i_1, i_2, \ldots, i_d}, \quad \mathbf{a}^{\mathbf{b}} = \prod_{i=1}^{d} a_i^{b_i}, \tag{4.3}$$

where $\mathbf{N} = (N_1, .., N_d)$. We use the notation $\mathbf{a} < \mathbf{b}$ if $a_1 < b_1, a_2 < b_2, \ldots, a_d < b_d$ (analogously for the other inequalities). We write $\mathbf{a} - n$ to indicate the product $\prod_{i=1}^{d} (a_i - n)$, if $n \in \mathbb{R}$. By β we denote the Beta function (see (2.2)) and we also use the notation

$$\beta(\mathbf{a}, \mathbf{b}) = \prod_{i=1}^{d} \beta(a_i, b_i)$$

if $\mathbf{a} = (a_1, .., a_d)$ and $\mathbf{b} = (b_1, .., b_d)$ belong to $(0, \infty)^d$.

Let $d, q \geq 1$ be integer numbers. The Hermite sheet constitutes a multiparameter version of the process (2.6). Let $(W(x), x \in \mathbb{R}^d)$ be a Wiener sheet, characterized by its covariance (1.11). Let $\mathbf{H} = (H_1, \ldots, H_d) \in \left(\frac{1}{2}, 1\right)^d$. The d-parameter Hermite process (or the Hermite sheet) $\left(Z_t^{\mathbf{H},q,d}, t \in \mathbb{R}_+^d \right)$ of order $q \geq 1$ and with Hurst parameter $\mathbf{H}$ is defined, for every $t \in \mathbb{R}_+$, as

$$Z_t^{\mathbf{H},q,d} = c(\mathbf{H}, q, d) \int_{\mathbb{R}^{dq}} dW(y_1) \dots dW(y_q) \qquad (4.4)$$

$$\times \left[\int_{[0,t]} (s - y_1)_+^{-\left(\frac{1}{2} + \frac{1-\mathbf{H}}{q}\right)} \dots (s - y_q)_+^{-\left(\frac{1}{2} + \frac{1-\mathbf{H}}{q}\right)} ds \right]$$

where $x_+ = \max(x, 0)$ and

$$c(\mathbf{H}, q, d)^2 = \frac{1}{q!} \frac{\mathbf{H}(2\mathbf{H} - 1)}{\beta\left(\frac{1}{2} - \frac{1-\mathbf{H}}{q}, \frac{2-2\mathbf{H}}{q}\right)^q} = \frac{1}{q!} \prod_{j=1}^d \left(\frac{H_j(2H_j - 1)}{\beta\left(\frac{1}{2} - \frac{1-H_j}{q}, \frac{2-2H_j}{q}\right)^q} \right)$$

$$(4.5)$$

If $d = 1$, we retrieve the constant (2.5). The notation $(s - y_1)_+^{-\left(\frac{1}{2} + \frac{1-\mathbf{H}}{q}\right)}$ is understood as $\prod_{j=1}^d (s_j - y_{1,j})^{-\left(\frac{1}{2} - \frac{1-H_j}{q}\right)}$ if $s = (s_1, \dots, s_d)$ and $y_1 = (y_{1,1}, y_{1,2}, \dots, y_{1,d})$. The constant $c(\mathbf{H}, q, d)$ is chosen such that $\mathbf{E}(Z_{\mathbf{H},d}^{(q)}(t))^2 = t^{2\mathbf{H}}$ for every $t \in \mathbb{R}_+^d$, see below.

We can also write, for every $t \in \mathbb{R}_+^d$,

$$Z_t^{\mathbf{H},q,d} = I_q(L_t)$$

where I_q stands for the multiple integral of order q with respect to the Wiener sheet and, for $y_1, \dots, y_q \in \mathbb{R}^d$, the kernel L_t (actually, $L_t = L^{\mathbf{H},q,d}$, but we omit these indices in the notation) is given by

$$L_t(y_1, \dots, y_q) = c(\mathbf{H}, q, d) \left[\int_{[0,t]} (s - y_1)_+^{-\left(\frac{1}{2} + \frac{1-\mathbf{H}}{q}\right)} \dots (s - y_q)_+^{-\left(\frac{1}{2} + \frac{1-\mathbf{H}}{q}\right)} ds \right].$$

$$(4.6)$$

Let us see that the Hermite sheet is well-defined and compute its covariance.

Lemma 4.7 *For every $t \in [0, \infty)^d$, the function L_t given by (4.6) belongs to $L^2(\mathbb{R}^d)$. Moreover, for every $s, t \in \mathbb{R}_+^d$, we have*

$$\mathbf{E} Z_t^{\mathbf{H},q,d} Z_s^{\mathbf{H},q,d} = \frac{1}{2} \left(t^{2\mathbf{H}} + s^{2\mathbf{H}} - |t - s|^{2\mathbf{H}} \right) := R^{\mathbf{H}}(t, s). \qquad (4.7)$$

Proof By Lemma 2.2, for $u = (u_1, \dots, u_d)$, $v = (v_1, \dots, v_d)$, $a = (a_1, \dots, a_d)$ with $a_i < -\frac{1}{2}$, we notice that

$$\int_{\mathbb{R}^d} (u-y)^a_+ (v-y)^a_+ dy = \beta(-1-2\mathbf{a}, \mathbf{a}+1)|u-v|^{2a+1} \left(\prod_{j=1}^d |u_j - v_j|^{2a_j+1} \right).$$

(4.8)

Let us compute the covariance of the random field $Z^{H,q,d}$. For $s = (s_1, ..., s_d)$, $t = (t_1, ..., t_d) \in \mathbb{R}^d_+$, we have

$$\mathbf{E} Z^{\mathbf{H},q,d}_t Z^{\mathbf{H},q,d}_s = q! c(\mathbf{H}, q, d)^2 \int_{\mathbb{R}^{dq}} dy_1 .... dy_q$$

$$\times \int_{[0,t]} (u-y_1)^{-\left(\frac{1}{2}+\frac{1-\mathbf{H}}{q}\right)}_+ \ldots (u-y_q)^{-\left(\frac{1}{2}+\frac{1-\mathbf{H}}{q}\right)}_+ du$$

$$\times \int_{[0,s]} (v-y_1)^{-\left(\frac{1}{2}+\frac{1-\mathbf{H}}{q}\right)}_+ \ldots (v-y_q)^{-\left(\frac{1}{2}+\frac{1-\mathbf{H}}{q}\right)}_+ dv$$

and by using Fubini's theorem,

$$\mathbf{E} Z^{\mathbf{H},q,d}_t Z^{\mathbf{H},q,d}_s = q! c(\mathbf{H}, q, d)^2 \int_{[0,t]} du \int_{[0,s]} dv$$

$$\times \left(\int_{\mathbb{R}^d} (u-y)^{-\left(\frac{1}{2}+\frac{1-\mathbf{H}}{q}\right)}_+ (v-y)^{-\left(\frac{1}{2}+\frac{1-\mathbf{H}}{q}\right)}_+ dy \right)^q.$$

We compute

$$\int_{\mathbb{R}^d} (u-y)^{-\left(\frac{1}{2}+\frac{1-\mathbf{H}}{q}\right)}_+ (v-y)^{-\left(\frac{1}{2}+\frac{1-\mathbf{H}}{q}\right)}_+ dy$$

$$= \int_{\mathbb{R}^d} dy_1 .... dy_d \prod_{j=1}^d (u_j - y_j)^{-\left(\frac{1}{2}+\frac{1-H_j}{q}\right)}_+ \prod_{j=1}^d (v_j - y_j)^{-\left(\frac{1}{2}+\frac{1-H_j}{q}\right)}_+$$

$$= \prod_{j=1}^d \int_{\mathbb{R}} dy_j (u_j - y_j)^{-\left(\frac{1}{2}+\frac{1-H_j}{q}\right)}_+ (v_j - y_j)^{-\left(\frac{1}{2}+\frac{1-H_j}{q}\right)}_+$$

and by (4.8),

$$\int_{\mathbb{R}^d} (u-y)^{-\left(\frac{1}{2}+\frac{1-H_j}{q}\right)}_+ (v-y)^{-\left(\frac{1}{2}+\frac{1-H_j}{q}\right)}_+ dy$$

$$= \prod_{j=1}^d \beta \left(\frac{1}{2} - \frac{1-H_j}{q}, \frac{2H_j - 2}{q} \right) |u_j - v_j|^{\frac{2H_j-2}{q}}$$

$$= \beta \left(\frac{1}{2} - \frac{1-\mathbf{H}}{q}, \frac{2\mathbf{H}-2}{q} \right) |u-v|^{\frac{2\mathbf{H}-2}{q}}.$$

Thus, for $s, t \in \mathbb{R}_+^d$,

$$\mathbf{E} Z_t^{\mathbf{H},q,d} Z_s^{\mathbf{H},q,d} = q! c(\mathbf{H}, q, d)^2 \beta \left(\frac{1}{2} - \frac{1-\mathbf{H}}{q}, \frac{2\mathbf{H}-2}{q} \right)^q$$

$$\times \int_{[0,t]} du \int_{[0,s]} dv |u - v|^{2\mathbf{H}-2}$$

$$= \frac{1}{2} \left(t^{2\mathbf{H}} + s^{2\mathbf{H}} - |t - s|^{2\mathbf{H}} \right) = R^{\mathbf{H}}(t, s),$$

where we used (4.5). ∎

4.2 Basic Properties

We give some immediate distributional and trajectorial properties of the Hermite sheet. We start with the concept of self-similarity for multiparameter stochastic processes.

Definition 4.7 Let $(X(x), x \in \mathbb{R}^d)$ be a d-parameter stochastic process and let $H = (H_1, ..., H_d) \in (0, 1)^d$. We will say that the process X is self-similar of index H (or H-self-similar) if for every $h = (h_1, ..., h_d) \in (0, \infty)^d$, the stochastic process

$$\left(h^H X \left(\frac{x}{h} \right), x \in \mathbb{R}^d \right) = \left(h_1^{H_1} .... h_d^{H_d} X \left(\frac{x_1}{h_1}, ..., \frac{x_d}{h_d} \right), x = (x_1, .., x_d) \in \mathbb{R}^d \right)$$

has the same finite dimensional distributions as the process X.

Proposition 4.23 *The Hermite sheet $Z^{\mathbf{H},q,d}$ given by (4.4) is $\mathbf{H}$-self-similar.*

Proof We have, for $\mathbf{h} = (h_1, ..., h_d) \in (0, \infty)^d$ and $\mathbf{t} = (t_1, ..., t_d) \in [0, \infty)^d$,

$$\mathbf{h}^{\mathbf{H}} Z_{\frac{t}{\mathbf{h}}}^{\mathbf{H},q,d}$$

$$= c(\mathbf{H}, q, d) \mathbf{h}^{\mathbf{H}} \int_{[0, \frac{t}{\mathbf{h}}]} \int_{\mathbb{R}^{dq}} dW(y_1) ... dW(y_q)$$

$$\times \left[\int_{[0,t]} (s - y_1)_+^{-\left(\frac{1}{2} + \frac{1-\mathbf{H}}{q} \right)} ... (s - y_q)_+^{-\left(\frac{1}{2} + \frac{1-\mathbf{H}}{q} \right)} ds \right]$$

$$= c(\mathbf{H}, q, d) \mathbf{h}^{\mathbf{H}} (h_1 ... h_d)^{-1} \int_{\mathbb{R}^{dq}} dW(y_1) ... dW(y_q)$$

$$\times \left[\int_{[0,t]} (\mathbf{h}s - y_1)_+^{-\left(\frac{1}{2} + \frac{1-\mathbf{H}}{q} \right)} ... (\mathbf{h}s - y_q)_+^{-\left(\frac{1}{2} + \frac{1-\mathbf{H}}{q} \right)} ds \right],$$

via the change of variables $\tilde{s}_j = s_j \mathbf{h}$ for $j = 1, .., d$. Now, by using $y_j = \mathbf{h} \tilde{y}_j$ for $j = 1, ..., d$,

$$\mathbf{h}^{\mathbf{H}} Z_{\frac{t}{\mathbf{h}}}^{\mathbf{H},q,d} = c(\mathbf{H},q,d)\mathbf{h}^{\mathbf{H}}(h_1...h_d)^{-1}$$

$$\times \int_{\mathbb{R}^{dq}} \left[\int_{[0,t]} (\mathbf{h}s - \mathbf{h}y_1)_+^{-\left(\frac{1}{2}+\frac{1-H}{q}\right)} \cdots (\mathbf{h}s - \mathbf{h}y_q)_+^{-\left(\frac{1}{2}+\frac{1-H}{q}\right)} ds \right]$$

$$dW(\frac{y_1}{\mathbf{h}})\ldots dW(\frac{y_q}{\mathbf{h}})$$

$$\stackrel{(d)}{=} Z_{\mathbf{t}}^{\mathbf{H},q,d},$$

where the last line is obtained via the scaling property of the Hermite sheet. ∎

Recall that the multidimensional increment of a random field is defined by (1.12).

Definition 4.8 We say that a d-parameter stochastic process $(X(x), x \in \mathbb{R}^d)$ has stationary increments if for every $h = (h_1, ..., h_d) \in (0, \infty)^d$, the d-parameter process

$$\left(\Delta X_{[h,x+h]}, x \in \mathbb{R}^d\right)$$

has the same finite-dimensional distributions as $(X(x), x \in \mathbb{R}^d)$.

Proposition 4.24 *The Hermite sheet $Z^{\mathbf{H},q,d}$ has stationary increments.*

Proof The argument is similar to that in the proof of Proposition 2.2.1. For $h \in (0, \infty)^d$,

$$\Delta Z_{[h,t+h]}^{\mathbf{H},q,d} = c(\mathbf{H},q,d) \int_{\mathbb{R}^{dq}} dW(y_1)\ldots dW(y_q)$$

$$\times \left[\int_{[h,t+h]} (s - y_1)_+^{-\left(\frac{1}{2}+\frac{1-H}{q}\right)} \cdots (s - y_q)_+^{-\left(\frac{1}{2}+\frac{1-H}{q}\right)} ds \right]$$

$$= c(\mathbf{H},q,d) \int_{\mathbb{R}^{dq}} dW(y_1)\ldots dW(y_q)$$

$$\times \left[\int_{[0,t]} (s+h - y_1)_+^{-\left(\frac{1}{2}+\frac{1-H}{q}\right)} \cdots (s+h - y_q)_+^{-\left(\frac{1}{2}+\frac{1-H}{q}\right)} ds \right]$$

where we made the change of variables $\tilde{u}_i = u_i - h_i$ for $i = 1, ..., d$. So,

$$\Delta Z_{[h,t+h]}^{\mathbf{H},q,d} = c(\mathbf{H},q,d) \int_{\mathbb{R}^{dq}} dW(y_1 + h)\ldots dW(y_q + h)$$

$$\times \left[\int_{[0,t]} (s - y_1)_+^{-\left(\frac{1}{2}+\frac{1-H}{q}\right)} \cdots (s - y_q)_+^{-\left(\frac{1}{2}+\frac{1-H}{q}\right)} ds \right]$$

$$\stackrel{(d)}{=} Z_t^{\mathbf{H},q,d}$$

where we used the fact that the Wiener sheet has stationary (multidimensional) increments. ∎

To analyze the sample paths regularity of the Hermite sheet, we recall the multi-parameter version of the Kolmogorov continuity theorem (see e.g. [6]).

Theorem 4.4 *Let* $(X_t, t \in T)$ *be a d-parameter process, vanishing on the axis, with T a compact subset of* $\mathbb{R}^d$. *Suppose that there exist constants* $C, p > 0$ *and* $\beta_1, .., \beta_d > 1$ *such that*

$$\mathbf{E}\left|\Delta X_{[t,t+h]}\right|^p \leq Ch_1^{\beta_1} \ldots h_d^{\beta_d}$$

for every $h = (h_1, \ldots, h_d) \in (0, \infty)^d$ *and for every* $t \in T$ *such that* $t + h \in T$. *Then X admits a continuous modification* $\tilde{X}$. *Moreover* $\tilde{X}$ *has Hölder continuous paths of any orders* $\delta = (\delta_1, \ldots, \delta_d)$ *with* $\delta_i \in (0, \frac{\beta_i - 1}{p})$ *for* $i = 1, \ldots, d$ *in the following sense: for every* $\omega \in \Omega$, *there exists* $C_\omega > 0$ *such that for every* $t, t' \in T$

$$\left|\Delta X_{[t,t']}(\omega)\right| \leq C_\omega |t - t'|^\delta$$

where $|t - t'|^\delta = \prod_{j=1}^d |t_j - t'_j|^{\delta_j}$ *if* $t = (t_1, \ldots, t_d), t' = (t'_1, \ldots, t'_d)$, *see convention (4.3).*

Proposition 4.25 *Let* $\left(Z_t^{\mathbf{H},q,d}, t \in \mathbb{R}^d\right)$ *be a Hermite sheet. Then for every* $s = (s_1, \ldots, s_d), t = (t_1, \ldots, t_d) \in \mathbb{R}_+^d$, *and for every* $p \geq 2$,

$$\mathbf{E}\left|\Delta Z_{[s,t]}^{\mathbf{H},q,d}\right|^p = \mathbf{E}|Z_1^{\mathbf{H},q,d}|^p |t_1 - s_1|^{H_1 p} \ldots |t_d - s_d|^{H_d p}.$$

In particular, the trajectories of the Hermite sheet $Z^{\mathbf{H},q,d}$ *are (modulo a modification), Hölder continuous of order* $\delta = (\delta_1, \ldots, \delta_d)$ *for every* $\delta_i \in (0, H_i)$ *for* $i = 1, \ldots, d$.

Proof By using the stationarity of the increments and the self-similarity of the Hermite sheet (Propositions 4.23 and 4.24), we can write

$$\mathbf{E}\left|\Delta Z_{[s,t]}^{\mathbf{H},q,d}\right|^p = \mathbf{E}\left|\Delta Z_{[0,t-s]}^{\mathbf{H},q,d}\right|^p = \mathbf{E}\left|(t-s)^{\mathbf{H}p} Z_1^{\mathbf{H},q,d}\right|^p$$
$$= \mathbf{E}\left|(t_1 - s_1)^{H_1 p} \ldots (t_d - s_d)^{H_d p} Z_1^{\mathbf{H},q,d}\right|^p$$
$$= \mathbf{E}|Z_1^{\mathbf{H},q,d}|^p |t_1 - s_1|^{H_1 p} \ldots |t_d - s_d|^{H_d p}.$$

The Hölder continuity follows from Theorem 4.4. ∎

4.3 Wiener Integral with Respect to the Hermite Sheet

The construction of the Wiener integral with respect to the d-parameter Hermite sheet follows the lines of the one-dimensional case described in Sect. 3.1. Let $(Z_t^{\mathbf{H},q,d}, t \in \mathbb{R}^d)$ be a d-parameter Hermite process defined by (4.4) and let $\mathcal{E}$ be the set of elementary functions on $\mathbb{R}^d$ of the form

$$f(t) = \sum_{k=1}^{N} a_k 1_{[t_k, t_{k+1})}(t) \tag{4.9}$$

with $a_k \in \mathbb{R}$ and $0 \le t_k < t_{k+1}$ for $j = 1, ..., N$. If f is as in (4.9), then we set

$$\int_{\mathbb{R}^d} f(u) dZ_u^{\mathbf{H},q,d} = \sum_{k=1}^{N} a_k \Delta Z_{[t_k, t_{k+1}]}^{\mathbf{H},q,d}, \tag{4.10}$$

where $\Delta Z_{[t_k, t_{k+1}]}^{H,q,d}$ stands for the high order increment over the interval $[t_k, t_{k+1}]$, see (1.12). Let us introduce the transfer operator

$$(Jf)(y_1, ..., y_q) = c(\mathbf{H}, q, d) \int_{\mathbb{R}^d} f(u) \prod_{j=1}^{q} (u - y_j)_+^{-\left(\frac{1}{2} + \frac{1-\mathbf{H}}{q}\right)} du, \tag{4.11}$$

for $y_1, ..., y_q \in \mathbb{R}^d$, with $c(\mathbf{H}, q, d)$ given by (4.5). Using the convention (4.3), this can be written as

$$(Jf)(y_1, ..., y_q) = c(\mathbf{H}, q, d) \int_{\mathbb{R}^d} f(u_1, ..., u_d) \prod_{j=1}^{q} \prod_{k=1}^{d} (u_k - y_{j,k})_+^{-\left(\frac{1}{2} + \frac{1-H_k}{q}\right)} du_1...du_d$$

if $y_j = (y_{j,1}, ..., y_{j,d})$ for $j = 1, ..., q$. By (4.10) and (1.12), we can write

$$\int_{\mathbb{R}^d} f(u) dZ_u^{\mathbf{H},q,d} = \sum_{k=1}^{N} a_k \Delta Z_{[t_k, t_{k+1}]}^{\mathbf{H},q,d}$$

$$= \sum_{k=1}^{N} a_k \left(\sum_{r \in \{0,1\}^d} (-1)^{d - \sum_{i=1}^{d} r_i} Z_{t_{k+r_1,1}, ..., t_{k+r_d,d}}^{\mathbf{H},q,d} \right)$$

if $t_k = (t_{k,1}, ..., t_{k,d})$ for $k = 1, ..., N + 1$. By the definition of the transfer operator (4.11), we get

$$\int_{\mathbb{R}^d} f(u) dZ_u^{\mathbf{H},q,d}$$

$$= \sum_{k=1}^{N} a_k \sum_{r \in \{0,1\}^d} (-1)^{d - \sum_{i=1}^{d} r_i}$$

$$\times \int_{\mathbb{R}^{dq}} J\left(1_{[0,t_{k+r_1,1}] \times ... \times [0,t_{k+r_d,d}]}\right)(y_1, ..., y_q) dW(y_1)....dW(y_q)$$

$$= \sum_{k=1}^{N} a_k \int_{\mathbb{R}^{dq}} J\left(1_{[t_{1,k}, t_{1,k+1}]} \times .... \times 1_{[t_{d,k}, t_{d,k+1}]}\right)(y_1, ..., y_q) dW(y_1)....dW(y_q)$$

$$= \int_{\mathbb{R}^{dq}} (Jf)(y_1, ..., y_q) dW(y_1)....dW(y_q).$$

This motivates the following definition: let $f : \mathbb{R}^d \to \mathbb{R}$ be a measurable function such that $f \in \mathcal{H}_{\mathbf{H},d}$, i.e.

$$\int_{\mathbb{R}^{dq}} (Jf)^2(y_1, ..., y_q) dy_1....dy_q < \infty.$$

Then we set by definition

$$\int_{\mathbb{R}^d} f(u) dZ_u^{\mathbf{H},q,d} := \int_{\mathbb{R}^{dq}} (Jf)(y_1, ..., y_q) dW(y_1)....dW(y_q) \tag{4.12}$$

with Jf given by (4.11). This will be called the Wiener integral of f with respect to Hermite sheet $Z^{\mathbf{H},q,d}$, or, as in the one-dimensional case, the Hermite-Wiener integral. In particular, this integral is well defined if

$$f \in |\mathcal{H}_{\mathbf{H},d}|$$

where $|\mathcal{H}_{\mathbf{H},d}|$ is the set of measurable function $f : \mathbb{R}^d \to \mathbb{R}$ such that

$$\|f\|^2_{|\mathcal{H}_{\mathbf{H},d}|} := \int_{\mathbb{R}^d} \int_{\mathbb{R}^d} |f(u)| \cdot |f(v)| |u - v|^{2\mathbf{H}-2} du dv < \infty. \tag{4.13}$$

The set $|\mathcal{H}_{\mathbf{H},d}|$ is is a Banach space with respect to the norm $\| \cdot \|_{|\mathcal{H}_{\mathbf{H},d}|}$ defined by (4.13) and we have the following inclusions (see [32]),

$$L^2(\mathbb{R}^d \cap L^1(\mathbb{R}^d)) \subset L^{\frac{1}{\mathbf{H}}}(\mathbb{R}^d) \subset |\mathcal{H}_{\mathbf{H},d}| \subset \mathcal{H}_{\mathbf{H},d}.$$

The Wiener integral with respect to the Hermite sheet is an isometry, Indeed, assume that f is a step function as in (4.9). Then

$$\mathbf{E} \left(\int_{\mathbb{R}^d} f(u) dZ_u^{\mathbf{H},q,d} \right)^2$$

$$= \sum_{k,l=1}^{N} a_k a_l \mathbf{E} \Delta Z_{[t_k,t_{k+1}]}^{\mathbf{H},q,d} \Delta Z_{[t_l,t_{l+1}]}^{\mathbf{H},q,d}$$

$$= \sum_{k,l=1}^{N} a_k a_l \sum_{r \in \{0,1\}^d} (-1)^{d-\sum_{i=1}^d r_i} \sum_{\rho \in \{0,1\}^d} (-1)^{d-\sum_{j=1}^d \rho_j} \mathbf{E} Z_{t_{k+r}}^{\mathbf{H},q,d} Z_{t_{l+\rho}}^{\mathbf{H},q,d}$$

$$= \sum_{k,l=1}^{N} a_k a_l \sum_{r \in \{0,1\}^d} (-1)^{d-\sum_{i=1}^d r_i} \sum_{\rho \in \{0,1\}^d} (-1)^{d-\sum_{j=1}^d \rho_j} R_{\mathbf{H}}(t_{k+r}, t_{l+\rho})$$

with $R^{\mathbf{H}}$ given by (4.7). By Lemma 2.1,

$$
\mathbf{E}\left(\int_{\mathbb{R}^d} f(u)dZ_u^{\mathbf{H},q,d}\right)^2
$$

$$
= \sum_{k,l=1}^{N} a_k a_l H_1(2H_1 - 1)\ldots H_d(2H_d - 1)
$$

$$
\int_{t_{k,1}}^{t_{k+1,1}} \ldots \int_{t_{k,d}}^{t_{k+1,d}} du_1\ldots du_d \int_{t_{l,1}}^{t_{l+1,1}} \ldots \int_{t_{l,d}}^{t_{l+1,d}} dv_1\ldots dv_d
$$

$$
\times |u_1 - v_1|^{2H_1-2} \ldots |u_d - v_d|^{2H_d-2}. \tag{4.14}
$$

Let us introduce the following notation, for $f, g \in \left|\mathcal{H}_{\mathbf{H},d}\right|$,

$$
\langle f, g\rangle_{\mathcal{H}_{\mathbf{H},d}} = \mathbf{H}(2\mathbf{H} - 1) \int_{\mathbb{R}^d} \int_{\mathbb{R}^d} f(u)g(v)|u - v|^{2\mathbf{H}-2} du dv
$$

and

$$
\|f\|_{\mathcal{H}_{\mathbf{H},d}}^2 = \mathbf{H}(2\mathbf{H} - 1) \int_{\mathbb{R}^d} \int_{\mathbb{R}^d} f(u)f(v)|u - v|^{2\mathbf{H}-2} du dv.
$$

Then, by (4.14),

$$
\mathbf{E}\left(\int_{\mathbb{R}^d} f(u)dZ_u^{\mathbf{H},q,d}\right)^2
$$

$$
= \sum_{k,l=1}^{N} a_k a_l \langle 1_{[t_{k,1},t_{k+1,1}]\times\ldots\times[t_{k,d},t_{k+1,d}]}, 1_{[t_{l,1},t_{l+1,1}]\times\ldots\times[t_{l,d},t_{l+1,d}]}\rangle_{\mathcal{H}_{\mathbf{H},d}}
$$

$$
= \sum_{k,l=1}^{N} a_k a_l \langle 1_{[t_k,t_{k+1}]}, 1_{[t_k,t_{k+1}]}\rangle_{\mathcal{H}_{\mathbf{H},d}}
$$

$$
= \langle f, f\rangle_{\mathcal{H}_{\mathbf{H},d}}.
$$

Since the set $\mathcal{E}$ is dense in $\left|\mathcal{H}_{\mathbf{H},d}\right|$ (see [12, 32]), the above isometry can be extended to $\left|\mathcal{H}_{\mathbf{H},d}\right|$. Therefore, for every $f, g \in \left|\mathcal{H}_{\mathbf{H},d}\right|$ we will have

$$
\mathbf{E}\left(\int_{\mathbb{R}^d} f(u)dZ_u^{\mathbf{H},q,d}\right)\left(\int_{\mathbb{R}^d} g(u)dZ_u^{\mathbf{H},q,d}\right) = \langle f, g\rangle_{\mathcal{H}_{\mathbf{H},d}}. \tag{4.15}
$$

4.4 The Stochastic Heat Equation with Hermite Noise

The construction of the Wiener integral with respect to a Hermite sheet allows to consider SPDEs (stochastic partial differential equations) with Hermite noise. We will illustrate here the case of the stochastic heat equation (but other types of SPDEs can also be treated, see e.g. [12] for the wave equation with Hermite noise).

The simplest deterministic heat equation in an open set $U \subset \mathbb{R}^d (d \geq 1)$ can be written as

$$\frac{\partial u}{\partial t} = \Delta u(t, x), \quad t \geq 0, x \in U \quad (4.16)$$

where Δ is the standard Laplacian operator, i.e. if $F : U \subset \mathbb{R}^d \to \mathbb{R}$, then

$$\Delta F(x) = \sum_{i=1}^{d} \frac{\partial^2 F}{\partial x_i^2}(x), \text{ if } x = (x_1, ..., x_d) \in U.$$

The Eq. (4.16) describes the heat flow in an homogeneous medium, $u(t, x)$ modelling the temperature at time t and at the point x in space. The stochastic heat equation constitutes a model for the heat flow under a random perturbation. In most situations, the random perturbation (or the random noise) is assumed to be a Gaussian field, the most prominent example being the space-time white noise (see e.g. [14] or [51]). Here we will analyze the stochastic heat equation with a non-Gaussian noise, namely the Hermite noise. More precisely, we consider the stochastic partial differential equation

$$\frac{\partial}{\partial t} u(t, x) = \Delta u(t, x) + \dot{Z}_{t,x}^{(H_0, \mathbf{H}), q, d+1}, \quad t \geq 0, x \in \mathbb{R}^d \quad (4.17)$$

with vanishing initial condition

$$u(0, x) = 0 \text{ for every } x \in \mathbb{R}^d.$$

In (4.17), $\dot{Z}^{(H_0, \mathbf{H}), q, d+1}$ stands for the formal derivative of

$$\left(Z_t^{(H_0, \mathbf{H}), q, d+1}, t \in \mathbb{R}^{d+1} \right),$$

which is a $d + 1$-parameter Hermite sheet, of order $q \geq 1$, with Hurst parameter $(H_0, \mathbf{H}) \in \left(\frac{1}{2}, 1\right)^{d+1}$, where $\mathbf{H} = (H_1, ..., H_d) \in \left(\frac{1}{2}, 1\right)^d$.

4.4.1 Existence of the Solution

Let us introduce the kernel

$$
G(t, x) = \begin{cases} (2\pi t)^{-d/2} \exp\left(-\frac{\|x\|^2}{2t}\right) & \text{if } t > 0, x \in \mathbb{R}^d, \\ 0 & \text{if } t \leq 0, x \in \mathbb{R}^d \end{cases}, \tag{4.18}
$$

where $\| \cdot \|$ denotes the Euclidean norm in $\mathbb{R}^d$. That is, G is the Green kernel (or the fundamental solution) associated to nthe heat equation, i.e it solves $\frac{\partial u}{\partial t} - \Delta u = 0$.

To define the solution to (4.17), we start with the following remark: if $f \in C^{1,2}(\mathbb{R}_+ \times \mathbb{R})$, set

$$
v(t, x) = \int_0^t ds \int_{\mathbb{R}^d} dy\, G(t - s, x - y) f(s, y). \tag{4.19}
$$

Then it is possible to prove that v defined by (4.19) solves the partial differential equation

$$
\frac{\partial}{\partial t} v(t, x) = \frac{1}{2} \Delta v(t, x) + f(t, x), \quad t \geq 0, x \in \mathbb{R}^d.
$$

This motivates the definition of the solution to (4.17): we replace $f(s, y) ds dy$ by $\dot{Z}_{s,y}^{(H_0, \mathbf{H}), q, d+1} ds dy$ which is interpreted as $dZ_{s,y}^{(H_0, \mathbf{H}), q, d+1}$, the Wiener integral with respect to the Hermite sheet introduced above in Sect. 4.3. More exactly, one defines the mild solution to (4.17) by

$$
u(t, x) = \int_0^t \int_{\mathbb{R}^d} G(t - s, x - y) dZ_{s,y}^{(H_0, \mathbf{H}), q, d+1}, \quad t \geq 0, x \in \mathbb{R}^d, \tag{4.20}
$$

where the above integral is a Hermite-Wiener integral. Using the definition (4.12) of the Hermite-Wiener integral, we cal also write

$$
u(t, x) = I_q(H_{t,x})
$$

where I_q is the multiple stochastic integral with respect to the Wiener sheet W and for $s_1 ..., s_q \in \mathbb{R}_+$ and $z_1, ..., z_q \in \mathbb{R}^d$,

$$
H_{t,x}\left((s_1, z_1), ..., (s_q, z_q)\right) = c(\mathbf{H}, q, d) \int_0^t du \int_{\mathbb{R}^d} dy\, G(t - u, x - y)
$$
$$
(u - s_1)^{-\left(\frac{1}{2} + \frac{1-H_0}{q}\right)} \ldots (u - s_q)^{-\left(\frac{1}{2} + \frac{1-H_0}{q}\right)}
$$
$$
(y - z_1)^{-\left(\frac{1}{2} + \frac{1-H}{q}\right)} \ldots (y - z_q)^{-\left(\frac{1}{2} + \frac{1-H}{q}\right)}. \tag{4.21}
$$

Let $\mathcal{S}(\mathbb{R}^d)$ be the Schwarz space of rapidly decreasing C^∞-functions on $\mathbb{R}^d$. For $f \in \mathcal{S}(\mathbb{R}^d)$, we define its Fourier transform by

$$(\mathcal{F}f)(\xi) = \int_{\mathbb{R}^d} e^{i\langle x,\xi\rangle}dx, \quad \xi \in \mathbb{R}^d,$$

with $\langle\cdot,\cdot\rangle$ standing for the Euclidean scalar product in $\mathbb{R}^d$. The Fourier transform of f is also well-defined when $f \in L^1(\mathbb{R}^d)$ or $f \in L^2(\mathbb{R}^d)$. We will use the Parseval identity

$$\alpha_{\mathbf{H}} \int_{\mathbb{R}^d}\int_{\mathbb{R}^d} \varphi(x)\psi(y)|x-y|^{2H-2}dydx = a_{\mathbf{H}} \int_{\mathbb{R}^d} \mathcal{F}\varphi(\xi)\overline{\mathcal{F}\psi}(\xi)|\xi|^{1-2H}d\xi \quad (4.22)$$

with some positive constant $a_{\mathbf{H}}$, for every φ, ψ such that

$$\int_{\mathbb{R}^d}\int_{\mathbb{R}^d} |\varphi(x)\psi(y)||x-y|^{2H-2}dydx < \infty.$$

We recall that the quantities $|\xi|^{1-2H}$ or $|x-y|^{2H-2}$ are understood via the convention (4.3).

Let us give a necessary and sufficient condition for the existence of the mild solution (4.20).

Proposition 4.26 *Let $T > 0$ and let $t \in [0, T]$, $x \in \mathbb{R}^d$ be fixed. Then the mapping*

$$(s, y) \rightarrow G(t - s, x - y)1_{(0,t)}(s)$$

belongs to $\left|\mathcal{H}_{(H_0,\mathbf{H}),d+1}\right|$ *if and only if*

$$d < 4H_0 + \sum_{i=1}^{d}(2H_i - 1). \quad (4.23)$$

In this case,

$$\sup_{t\in[0,T]} \mathbf{E}u(t, x)^2 < \infty.$$

Proof We have, by using the isometry of the Hermite-Wiener integral (4.15)

$$\mathbf{E}u(t, \mathbf{x})^2 = \|G(t - \cdot, \mathbf{x} - *)1_{(0,t)}(\cdot)\|^2_{\mathcal{H}_{\mathbf{H},d+1}}$$

$$= \alpha_{H_0} \int_0^t du \int_0^t dv|u - v|^{2H_0-2} \int_{\mathbb{R}^d}\int_{\mathbb{R}^d} dydz$$

$$G(t - u, \mathbf{x} - \mathbf{y})G(t - v, \mathbf{x} - \mathbf{z})f(\mathbf{y} - \mathbf{z})$$

where G is defined by (4.18) and

$$f(\mathbf{y} - \mathbf{z}) = \prod_{i=1}^{d} H_i(2H_i - 1)|y_i - z_i|^{2H_i-2} = \mathbf{H}(2\mathbf{H} - 1)|\mathbf{y} - \mathbf{z}|^{2H-2}.$$

We recall that the Fourier transform of the Green kernel (4.18) with respect to the spatial variable is given by

$$\mathcal{F}G(t, \cdot)(\xi) = e^{-\frac{t\|\xi\|^2}{2}}. \tag{4.24}$$

By using (4.24) and the Parseval relation (4.22), we obtain, with $c(H_0, \mathbf{H})$ a generic constant that may change,

$$\mathbf{E}u(t, \mathbf{x})^2$$
$$= c(H_0, \mathbf{H}) \int_0^t du \int_0^t dv |u - v|^{2H_0 - 2} \int_{\mathbb{R}^d} d\xi \mathcal{F}G(t - u, \cdot)(\xi)\overline{\mathcal{F}G(t - u, \cdot)}(\xi)|\xi|^{1-2\mathbf{H}}$$
$$= c(H_0, \mathbf{H}) \int_0^t du \int_0^t dv |u - v|^{2H_0 - 2} \int_{\mathbb{R}^d} d\xi e^{-\frac{(t-u)\|\xi\|^2}{2}} e^{-\frac{(t-v)\|\xi\|^2}{2}} |\xi|^{1-2\mathbf{H}}$$

where

$$|\xi|^{1-2\mathbf{H}} = \prod_{j=1}^d |\xi_j|^{1-2H_j}.$$

Therefore, by the change of variables $\sqrt{u + v}\xi_i = \tilde{\xi}_i$ for $i = 1, ..., d$, we find

$$\mathbf{E}u(t, \mathbf{x})^2$$
$$= c(H_0, \mathbf{H}) \int_0^t du \int_0^t dv |u - v|^{2H_0 - 2} \int_{\mathbb{R}^d} d\xi e^{-\frac{1}{2}(u+v)\|\xi\|^2} \prod_{j=1}^d |\xi_j|^{1-2H_j}$$
$$= c(H_0, \mathbf{H}) \int_0^t du \int_0^t dv |u - v|^{2H_0 - 2}(u + v)^{-\frac{d}{2}}(u + v)^{H_1 + ... + H_d - d} \int_{\mathbb{R}^d} d\xi e^{-\frac{1}{2}\|\xi\|^2}$$
$$= c(H_0, \mathbf{H}) \int_0^t du \int_0^t dv |u - v|^{2H_0 - 2}(u + v)^{-\frac{d}{2}}(u + v)^{H_1 + ... + H_d - d},$$

where we used the fact that the integral $\int_{\mathbb{R}^d} d\xi e^{-\frac{1}{2}\|\xi\|^2}$ is finite. Next, by the change of variables $\frac{v}{u} = z$ in the integral dv,

$$\mathbf{E}u(t, \mathbf{x})^2$$
$$= c(H_0, \mathbf{H}) \int_0^t du \int_0^u dv |u - v|^{2H_0 - 2}(u + v)^{-\frac{d}{2}}(u + v)^{H_1 + ... + H_d - d}$$
$$= c(H_0, \mathbf{H}) \int_0^t du \times u^{2H_0 - 1}u^{H_1 + ... + H_d - d} \int_0^1 dz(1 - z)^{2H_0 - 2}(1 + z)^{H_1 + ... + H_d - d}$$
$$= c(H_0, \mathbf{H}) \int_0^t du \times u^{2H_0 - 1}u^{H_1 + ... + H_d - d}$$

since the integral $\int_0^1 dz(1-z)^{2H_0-2}(1+z)^{H_1+...+H_d-d}$ is finite for $H_0 > \frac{1}{2}$. We finally notice that the integral $\int_0^t duu^{2H_0-1}u^{H_1+...+H_d-d}$ converges if and only if $d < 2H_0 + H_1 + ... + H_d$, and this leads to (4.23). ∎

4.4.2 Self-similarity

We prove that the solution (4.20) is self-similar with respect to its temporal variable.

Proposition 4.27 *Assume (4.23) and let* $x \in \mathbb{R}^d$ *be fixed. Then the process* $(u(t, x),$ $t \geq 0)$ *is self-similar of order*

$$\gamma = H_0 - \frac{d}{2} + \frac{H_1 + ... + H_d}{2}. \tag{4.25}$$

Proof Let $a > 0$. By (4.21),

$$u(at, x)$$
$$= c(\mathbf{H}, q, d) \int_{\mathbb{R}^{dq}} W(ds_1, dz_1)....W(ds_q, dz_q) \int_0^{at} du \int_{\mathbb{R}^d} dyG(at - u, x - y)$$
$$(u - s_1)^{-\left(\frac{1}{2}+\frac{1-H_0}{q}\right)} \ldots (u - s_q)^{-\left(\frac{1}{2}+\frac{1-H_0}{q}\right)}$$
$$(y - z_1)^{-\left(\frac{1}{2}+\frac{1-\mathbf{H}}{q}\right)} \ldots (y - z_q)^{-\left(\frac{1}{2}+\frac{1-\mathbf{H}}{q}\right)}$$
$$= c(\mathbf{H}, q, d)a \int_{\mathbb{R}} \int_{\mathbb{R}^d} \cdots \int_{\mathbb{R}} \int_{\mathbb{R}^d} W(ds_1, dz_1) \ldots W(ds_q, dz_q)$$
$$\int_0^t du \int_{\mathbb{R}^d} dyG(a(t - u), x - y)(au - s_1)_+^{-\left(\frac{1}{2}+\frac{1-H_0}{q}\right)} \ldots (au - s_q)_+^{-\left(\frac{1}{2}+\frac{1-H_0}{q}\right)}$$
$$(y - z_1)^{-\left(\frac{1}{2}+\frac{1-\mathbf{H}}{q}\right)} \ldots (y - z_q)^{-\left(\frac{1}{2}+\frac{1-\mathbf{H}}{q}\right)}$$
$$= c(\mathbf{H}, q, d)a \int_{\mathbb{R}} \int_{\mathbb{R}^d} \cdots \int_{\mathbb{R}} \int_{\mathbb{R}^d} W(d(as_1), dz_1) \ldots W(d(as_q), dz_q)$$
$$\int_0^t du \int_{\mathbb{R}^d} dyG_\alpha(a(t - u), x - y)(au - as_1)_+^{-\left(\frac{1}{2}+\frac{1-H_0}{q}\right)} \ldots (au - as_q)_+^{-\left(\frac{1}{2}+\frac{1-H_0}{q}\right)}$$
$$(y - z_1)_+^{-\left(\frac{1}{2}+\frac{1-\mathbf{H}}{q}\right)} \ldots (y - z_q)_+^{-\left(\frac{1}{2}+\frac{1-\mathbf{H}}{q}\right)}.$$

Using the fact that the Brownian sheet W is $\frac{1}{2}$-self-similar with respect to its time variable and its increments are stationary in space, we get (we denote, as usual, by $\equiv^{(d)}$ the equivalence of the finite dimensional distributions),

$$u(at, x) \equiv^{(d)} c(\mathbf{H}, q, d) a a^{-q\left(\frac{1}{2} + \frac{1-H_0}{q}\right)} a^{\frac{q}{2}} \int_{\mathbb{R}} \int_{\mathbb{R}^d} \cdots \int_{\mathbb{R}} \int_{\mathbb{R}^d} W(ds_1, dz_1) \ldots W(ds_q, dz_q)$$

$$\int_0^t du \int_{\mathbb{R}^d} dy \, G(a(t-u), -y)(u-s_1)_+^{-\left(\frac{1}{2} + \frac{1-H_0}{q}\right)} \ldots (u-s_q)_+^{-\left(\frac{1}{2} + \frac{1-H_0}{q}\right)}$$

$$(y - z_1)_+^{-\left(\frac{1}{2} + \frac{1-H}{q}\right)} \ldots (y - z_q)_+^{-\left(\frac{1}{2} + \frac{1-H}{q}\right)}.$$

We use the following property of the Green kernel

$$G(at, x) = a^{-\frac{d}{2}} G(t, a^{-\frac{1}{2}} x).$$

Then

$$u(at, x) \equiv^{(d)} c(\mathbf{H}, q, d) a^{H_0} a^{-\frac{d}{2}} \int_{\mathbb{R}} \int_{\mathbb{R}^d} \cdots \int_{\mathbb{R}} \int_{\mathbb{R}^d} W(ds_1, dz_1) \ldots W(ds_q, dz_q)$$

$$\int_0^t du \int_{\mathbb{R}^d} dy \, G(t-u, -a^{-\frac{1}{2}} y)(u-s_1)_+^{-\left(\frac{1}{2} + \frac{1-H_0}{q}\right)} \ldots (u-s_q)_+^{-\left(\frac{1}{2} + \frac{1-H_0}{q}\right)}$$

$$(y - z_1)_+^{-\left(\frac{1}{2} + \frac{1-H}{q}\right)} \ldots (y - z_q)_+^{-\left(\frac{1}{2} + \frac{1-H}{q}\right)}.$$

Next, we make the change of variables $a^{-\frac{1}{2}} y_i = \tilde{y}_i$ for $i = 1, ..., d$ and then $\tilde{z}_i = a^{-\frac{1}{2}} z_i$ for $i = 1, ..., d$. We will obtain

$$u(at, x)$$

$$\equiv^{(d)} c(\mathbf{H}, q, d) a^{H_0} a^{-\frac{d}{2}} a^{\frac{d}{2}} \int_{\mathbb{R}} \int_{\mathbb{R}^d} \cdots \int_{\mathbb{R}} \int_{\mathbb{R}^d} W(ds_1, d(a^{\frac{1}{2}} z_1)) \ldots W(ds_q, d(a^{\frac{1}{2}} z_q))$$

$$\int_0^t du \int_{\mathbb{R}^d} dy \, G(t-u, -y)(u-s_1)_+^{-\left(\frac{1}{2} + \frac{1-H_0}{q}\right)} \ldots (u-s_q)_+^{-\left(\frac{1}{2} + \frac{1-H_0}{q}\right)}$$

$$(a^{\frac{1}{2}}(y - z_1))_+^{-\left(\frac{1}{2} + \frac{1-H}{q}\right)} \ldots (a^{\frac{1}{2}}(y - z_q))_+^{-\left(\frac{1}{2} + \frac{1-H}{q}\right)}$$

$$= c(\mathbf{H}, q, d) a^{H_0} a^{-\frac{q}{2} \sum_{i=1}^d \left(\frac{1}{2} + \frac{1-H_i}{q}\right)}$$

$$\int_{\mathbb{R}} \int_{\mathbb{R}^d} \cdots \int_{\mathbb{R}} \int_{\mathbb{R}^d} W(ds_1, d(a^{\frac{1}{2}} z_1)) \ldots W(ds_q, d(a^{\frac{1}{2}} z_q))$$

$$\int_0^t du \int_{\mathbb{R}^d} dy \, G(t-u, -y)(u-s_1)_+^{-\left(\frac{1}{2} + \frac{1-H_0}{q}\right)} \ldots (u-s_q)_+^{-\left(\frac{1}{2} + \frac{1-H_0}{q}\right)}$$

$$(y - z_1)_+^{-\left(\frac{1}{2} + \frac{1-H}{q}\right)} \ldots (y - z_q)_+^{-\left(\frac{1}{2} + \frac{1-H}{q}\right)}.$$

Finally, we use the spatial scaling property of the Brownian sheet W to get

$$u(at, x) \equiv^{(d)} c(\mathbf{H}, q, d)a^{H_0} a^{-\frac{q}{2a}\sum_{i=1}^{d}\left(\frac{1}{2}+\frac{1-H_i}{q}\right)}a^{\frac{dq}{4}}$$

$$\int_{\mathbb{R}}\int_{\mathbb{R}^d}\cdots\int_{\mathbb{R}}\int_{\mathbb{R}^d} W(ds_1, z_1)\ldots W(ds_q, z_q)$$

$$\int_0^t du \int_{\mathbb{R}^d} dy G(t-u, -y)(u-s_1)^{-\left(\frac{1}{2}+\frac{1-H_0}{q}\right)}\ldots(u-s_q)^{-\left(\frac{1}{2}+\frac{1-H_0}{q}\right)}$$

$$(y-z_1)^{-\left(\frac{1}{2}+\frac{1-\mathbf{H}}{q}\right)}\ldots(y-z_q)^{-\left(\frac{1}{2}+\frac{1-\mathbf{H}}{q}\right)}$$

so for $x \in \mathbb{R}$ fixed,

$$(u(at, x), t \geq 0) \equiv^{(d)} \left(a^{H_0 - \frac{d}{2} + \frac{\sum_{i=1}^{d} H_i}{2}} u(t, x), t \geq 0\right) = (a^\gamma u(t, x), t \geq 0)$$

with γ in (4.25). ∎

Remark 4.7 Notice that the self-similarity index γ given by (4.25) is strictly positive, due to condition (4.23). Also, we have $\gamma < H_0 < 1$ since

$$\gamma = H_0 - \frac{d - \sum_{i=1}^{d} H_i}{2} \leq H_0 < 1.$$

4.4.3 Regularity of Sample Paths

Let us study the regularity of the mapping $t \to u(t, x)$, with $x \in \mathbb{R}^d$ fixed. To this end, we start by analyzing the temporal increment of the solution.

Proposition 4.28 *Assume (4.23). Then for every $0 \leq s < t$ and for every $x \in \mathbb{R}^d$,*

$$\mathbf{E}\,|u(t, x) - u(s, x)|^2 \leq C|t - s|^{2\gamma}, \tag{4.26}$$

with γ given by (4.25) and with $C > 0$ not depending on s, t, x.

Proof We write, for $0 \leq s < t$ and for $x \in \mathbb{R}^d$,

$$u(t, x) - u(s, x) = \int_s^t \int_{\mathbb{R}^d} G(t-u, x-y)dZ^{\mathbf{H},q,d}$$

$$+ \int_0^s \int_{\mathbb{R}^d} (G(t-u, x-y) - G(s-u, x-y))\, dZ^{\mathbf{H},q,d}_{u,y}$$

$$=: T_1(s, t) + T_2(s, t).$$

The two terms $T_1(s, t)$ and $T_2(s, t)$ from above will be estimated separately. First, we have, by the isometry (3.6) and (4.22),

$$\mathbf{E}|T_1(s,t)|^2$$

$$= \alpha_{H_0} \int_s^t \int_s^t du\,dv|u-v|^{2H_0-2}$$

$$\int_{\mathbb{R}^d} \int_{\mathbb{R}^d} dy\,dz\, G(t-u,x-y)G(t-v,x-y)|y-z|^{2\mathbf{H}-2}$$

$$= c(H_0,\mathbf{H}) \int_s^t \int_s^t du\,dv|u-v|^{2H_0-2} \int_{\mathbb{R}^d} d\xi\, e^{-\frac{1}{2}(t-u+s-v)\|\xi\|^2} \prod_{j=1}^d |\xi_j|^{1-2H_j}$$

$$= c(H_0,\mathbf{H}) \int_0^{t-s} \int_0^{t-s} du\,dv|u-v|^{2H_0-2} \int_{\mathbb{R}^d} d\xi\, e^{-\frac{1}{2}(u+v)\|\xi\|^2} \prod_{j=1}^d |\xi_j|^{1-2H_j}.$$

We use the change of variables $\tilde{\xi}_j = \sqrt{u+v}\,\xi_j$ for $j=1,...,d$ to get

$$\mathbf{E}|T_1(s,t)|^2 = c(H_0,\mathbf{H}) \int_0^{t-s} \int_0^{t-s} du\,dv|u-v|^{2H_0-2}(u+v)^{-d+H_1+...+H_d}$$

$$\times \int_{\mathbb{R}^d} d\xi\, e^{-\frac{1}{2}\|\xi\|^2} \prod_{j=1}^d |\xi_j|^{1-2H_j}$$

$$= c(H_0,\mathbf{H}) \int_0^{t-s} \int_0^{t-s} du\,dv|u-v|^{2H_0-2}(u+v)^{-d+H_1+...+H_d}$$

$$= c(H_0,\mathbf{H}) \int_0^{t-s} dv\, u^{2\gamma-1} \int_0^1 dz(1-z)^{2H_0-2}(1+z)^{H_1+...+H_d-d}$$

$$= c(H_0,\mathbf{H})|t-s|^{2\gamma}$$

where γ is defined by (4.25). Next,

$$\mathbf{E}|T_2(s,t)|^2$$

$$= c(H_0,\mathbf{H}) \int_0^s \int_0^s du\,dv|u-v|^{2H_0-2} \int_{\mathbb{R}^d} d\xi|\xi|^{1-2\mathbf{H}}$$

$$\left(e^{-\frac{1}{2}(t-u)|\xi|^2} - e^{-\frac{1}{2}(s-u)|\xi|^2}\right)\left(e^{-(t-v)|\xi|^2} - e^{-(s-v)|\xi|^2}\right)$$

$$= c(H_0,\mathbf{H})|t-s|^{2H_0} \int_0^{\frac{s}{t-s}} \int_0^{\frac{s}{t-s}} du\,dv|u-v|^{2H_0-2} \int_{\mathbb{R}^d} d\xi|\xi|^{1-2\mathbf{H}}$$

$$\left(e^{-\frac{1}{2}(t-s)(1+u)|\xi|^2} - e^{-\frac{1}{2}(t-s)u|\xi|^2}\right)\left(e^{-\frac{1}{2}(t-s)(1+v)|\xi|^2} - e^{-\frac{1}{2}(t-s)v|\xi|^2}\right)$$

where we performed the change of variables $\tilde{u} = \frac{s-u}{t-s}$, $\tilde{v} = \frac{s-v}{t-s}$. in the next step, we set $\tilde{\xi}_i = (t-s)^{\frac{1}{2}}\xi_i$ for every $i=1,...,d$. In this way,

$$\mathbf{E}|T_2(s, t)|^2$$

$$= C|t - s|^{2H_0}(t - s)^{-\frac{d}{2}+\sum_{i=1}^{d}(1-2H_i)} \int_0^{\frac{s}{t-s}} \int_0^{\frac{s}{t-s}} dudv|u - v|^{2H_0-2} \int_{\mathbb{R}^d} d\xi|\xi|^{1-2H}$$

$$\left(e^{-\frac{1}{2}(2+u+v)|\xi|^2} - 2e^{-\frac{1}{2}(1+u+v)|\xi|^2} + e^{-\frac{1}{2}(u+v)|\xi|^2} - 2\right)$$

$$\leq C(t - s)^{2\gamma} \int_0^{\infty} \int_0^{\infty} dudv|u - v|^{2H_0-2} \int_{\mathbb{R}^d} d\xi|\xi|^{1-2H}$$

$$\left(e^{-\frac{1}{2}(2+u+v)|\xi|^2} - 2e^{-\frac{1}{2}(1+u+v)|\xi|^2} + e^{-\frac{1}{2}(u+v)|\xi|^2}\right)$$

$$= C(t - s)^{2\gamma} \int_{\mathbb{R}^d} d\xi|\xi|^{1-2H}e^{-\frac{1}{2}|\xi|^2} \int_0^{\infty} \int_0^{\infty} dudv|u - v|^{2H_0-2}$$

$$\left[(2 + u + v)^{-\frac{d}{2}-\frac{1}{2}\sum_{i=1}^{d}(1-2H_i)} - 2(1 + u + v)^{-\frac{d}{2}-\frac{1}{2}\sum_{i=1}^{d}(1-2H_i)}\right.$$

$$\left.+(u + v)^{-\frac{d}{2}-\frac{1}{2}\sum_{i=1}^{d}(1-2H_i)}\right]$$

$$= C(t - s)^{2\gamma} \int_0^{\infty} \int_0^{\infty} dudv|u - v|^{2H_0-2}$$

$$\left[(2 + u + v)^{2\gamma-2H_0} - 2(1 + u + v)^{2\gamma-2H_0} + (u + v)^{2\gamma-2H_0}\right].$$

For u, v close to infinity, we have

$$(2 + u + v)^{2\gamma-2H_0} - 2(1 + u + v)^{2\gamma-2H_0} + (u + v)^{2\gamma-2H_0} \leq C|u + v|^{2\gamma-2H_0-2}$$

and the above integral $dudv$ is convergent at infinity because $2\gamma < 2$ or equivalently

$$H_0 - \frac{1}{4}\left(d - \sum_{i=1}^{d} H_i\right) \leq 1.$$

∎

As an immediate consequence of Proposition 4.28, we obtain the Hölder regularity in time of the solution.

Corollary 4.2 *The mapping $t \to u(t, x)$ is Hölder continuous of order δ for every $\delta \in (0, \gamma)$, with γ given by (4.25).*

Proof From (4.26) and by using the hypercontractivity (1.31), we have for $p \geq 2$,

$$\mathbf{E}|u(t, x) - u(s, x)|^p \leq C_p|t - s|^{p\gamma}$$

for every $0 \leq s < t$ and $x \in \mathbb{R}^d$. The conclusion follows by Kolmogorov's continuity criterion. ∎

4.4.4 A Decomposition Theorem

Consider the random field $(u(t, x), t \geq 0, x \in \mathbb{R}^d)$ which is mild solution to (4.17) and assume (4.23) is satisfied. We have seen in Proposition 4.27 that the the random field u is self-similar in time and it is pretty obvious that it has no stationary increments with respect to the time variable. The purpose is to give a decomposition of the solution to the fractional stochastic heat equation (4.17) as a sum of a self-similar process in time with temporal stationary increments and of another process with very nice sample paths in time. This decomposition is useful, among others, to get the behavior of p-variation of the solution and to estimate the drift parameter of the stochastic heat equation.

We will use the so-called *pinned string method* introduced in [25] and then used by several authors (in e.g. [21, 26, 38, 45]). Let us set, for $t \geq 0, x \in \mathbb{R}^d$,

$$U(t, x) = \int_{\mathbb{R}} \int_{\mathbb{R}^d} (G((t - s)_+, x - y) - G((-s)_+, x - y)) \, dZ_{s,y}^{(H_0, \mathbf{H}), q, d+1}.$$

(4.27)

This is called the *pinned string process.* We can also write

$$U(t, x) = \int_0^t \int_{\mathbb{R}} G(t - s, x - y) dZ_{s,y}^{(H_0, \mathbf{H}), q, d+1}$$
$$+ \int_{-\infty}^0 \int_{\mathbb{R}} (G(t - s, x - y) - G(-s, x - y)) \, dZ_{s,y}^{(H_0, \mathbf{H}), q, d+1}.$$

We will first show that U has the scaling property in time and it also has stationary temporal increments. Then we will show that the difference $u(t, x) - U(t, x)$ has C^∞ -sample paths with respect to the time variable.

Proposition 4.29 *Let $x \in \mathbb{R}^d$ be fixed. Then the process $(U(t, x), t \geq 0)$ defined by (4.27) is γ-self-similar and it has stationary increments.*

Proof The self-similarity follows as above in the proof of Proposition 4.27. Let us show that for every $h > 0$, the stochastic processes

$$(U(t + h, x) - U(h, x), t \geq 0) \text{ and } (U(t, x)), t \geq 0)$$

have the same finite dimensional distributions. We can write, for $h > 0$,

$$U(t + h, x) - U(h, x)$$
$$= \int_{\mathbb{R}} \int_{\mathbb{R}^d} (G((t + h - s)_+, x - y) - G((h - s)_+, x - y)) \, dZ_{s,y}^{(H_0, \mathbf{H}), q, d+1}$$
$$\stackrel{(d)}{=} \int_{\mathbb{R}} \int_{\mathbb{R}^d} (G((t - s)_+, x - y) - G((-s)_+, x - y)) \, dZ_{s,y}^{(H_0, \mathbf{H}), q, d+1}$$
$$= U(t, x)$$

where we used the fact that the process $(Z^{(H_0,\mathbf{H}),q,d+1}(t,x), t \geq 0)$ has stationary increments in time. ∎

From the above result, it follows (see e.g. [46]) that the covariance of the process $(U(t,x), t \geq 0)$ is given by

$$EU(t,x)U(s,x) = \frac{EU(1,x)^2}{2}\left(t^{2\gamma} + s^{2\gamma} - |t-s|^{2\gamma}\right), \quad t,s \geq 0.$$

On the other hand, since $(U(t,x), t \geq 0)$ is not a Gaussian process, the covariance did not determine the probability distribution of this stochastic process (except when $d = 1$).

Set, for $t > 0$ and $x \in \mathbb{R}^d$,

$$Y(t,x) = u(t,x) - U(t,x)$$

so

$$Y(t,x) = -\int_{-\infty}^0 \int_{\mathbb{R}^d} (G(t-s, x-y) - G(-s, x-y))\, dZ_{s,y}^{(H_0,\mathbf{H}),q,d+1}. \quad (4.28)$$

We will show that the random field Y has smooth sample paths with respect to the time variable.

Proposition 4.30 *Let* $(Y(t,x), t > 0, x \in \mathbb{R}^d)$ *be given by (4.28) and assume (4.23). Then for every* $x \in \mathbb{R}^d$, *the sample paths* $t \to Y(t,x)$ *are absolutely continuous and of class* C^∞ *on* $(0, \infty)$.

Proof Set

$$Y'(t,x) = -\int_{-\infty}^0 \int_{\mathbb{R}^d} \frac{\partial}{\partial t} G(t-s, x-y) dZ_{s,y}^{(H_0,\mathbf{H}),q,d+1}, \quad t > 0, x \in \mathbb{R}^d, \quad (4.29)$$

the formal derivative of Y with respect to the time variable t. We will have (again $C > 0$ denotes a generic constant) via (4.22)

$$E|Y'(t,x)|^2 = C \int_{-\infty}^0 \int_{-\infty}^0 du dv |u-v|^{2H_0-2} \int_{\mathbb{R}^d} d\xi |\xi|^{1-2\mathbf{H}}$$

$$\left(\frac{\partial}{\partial t} e^{-(t-u)\|\xi\|^2}\right)\left(\frac{\partial}{\partial t} e^{-(t-v)\|\xi\|^2}\right)$$

$$= C \int_{-\infty}^0 \int_{-\infty}^0 du dv |u-v|^{2H_0-2} \int_{\mathbb{R}^d} d\xi |\xi|^{1-2\mathbf{H}+4}$$

$$\times e^{-(t-u)\|\xi\|^2} e^{-(t-v)\|\xi\|^2}$$

$$= C \int_t^\infty \int_t^\infty du dv |u-v|^{2H_0-2} \int_{\mathbb{R}^d} d\xi |\xi|^{5-2\mathbf{H}} e^{-(u+v)\|\xi\|^2}$$

and by setting $(u + v)^{\frac{1}{2}}\xi_i = \tilde{\xi}_i$ for $i = 1, ..., d$, we get

$$\mathbf{E}|Y'(t, x)|^2 = C \int_t^\infty \int_t^\infty du dv |u - v|^{2H_0 - 2} (u + v)^{-\frac{d}{2}} (u + v)^{-\frac{1}{2} \sum_{i=1}^d (5 - 2H_i)}$$
$$\int_{\mathbb{R}^d} d\xi |\xi|^{1 - 2H + 2\alpha} e^{-|\xi|^2}$$
$$= C \int_t^\infty \int_t^\infty du dv |u - v|^{2H_0 - 2} (u + v)^{-\frac{d}{2}} (u + v)^{-\frac{1}{2} \sum_{i=1}^d (5 - 2H_i)}.$$

Thus, with $z = \frac{u}{v}$

$$\mathbf{E}|Y'(t, x)|^2$$
$$= C \int_t^\infty dv \int_v^\infty du (u - v)^{2H_0 - 2} (u + v)^{-\frac{2d}{2} - 2d + \sum_{i=1}^d H_i}$$
$$= C \int_t^\infty dv v^{2H_0 - 1 - \frac{2d}{2} - 2d + \sum_{i=1}^d H_i} \int_1^\infty dz (z - 1)^{2H_0 - 1} (1 + z)^{-\frac{2d}{\alpha} - 2d + \frac{2}{\alpha} \sum_{i=1}^d H_i}$$
$$= C \int_t^\infty dv v^{2\gamma - 1 - 2d} \int_1^\infty (1 - z)^{2H_0 - 2} (1 + z)^{2\gamma - 2H_0 - 2d}.$$

The integral dv is finite because $2\gamma - 2d < 0$ (see Remark 4.7) and the integral dx is finite at 1 because $2H_0 > 1$ and at infinity because $2\gamma - 2d - 1 < 0$ (again by Remark 4.7). Therefore $(Y'(t, x), t > 0)$ is a well defined random field and consequently $t \to Y(t, x)$ is absolute continuous and of class C^1 on $(0, \infty)$. Similarly (see [26, 38] or [45] for details), we can deal with the nth derivative and we can show that $t \to Y(t, x)$ is of class C^∞ on $(0, \infty)$. ∎

4.4.5 p-Variation

We will use the above decomposition theorem in order to obtain the p-variation in time of the solution u. Let us first define the concept of p-variation. Consider $0 \le A_1 < A_2$ two real numbers and let

$$t_i = A_1 + \frac{i}{N}(A_2 - A_1), \quad i = 0, ..., N \tag{4.30}$$

be a partition of the interval $[A_1, A_2]$. Let $(v(t, x), t \ge 0, x \in \mathbb{R}^d)$ a general random field and define, for $x \in \mathbb{R}^d$, $p > 0$ and $N \ge 1$

$$S_{[A_1, A_2]}^{N, p}(v(\cdot, x)) = \sum_{i=0}^{N-1} |v(t_{i+1}, x) - v(t_i, x)|^p. \tag{4.31}$$

We will say that v admits a temporal p-variation over the interval $[A_1, A_2]$ if the sequence $\left(S^{N,p}_{[A_1,A_2]}(v(\cdot, x)), N \geq 1 \right)$ converges in probability as $N \to \infty$.

For the solution to the fractional stochastic heat equation, we have the following result. Recall that γ is given by (4.25).

Theorem 4.5 *Let* $(u(t, x), t \geq 0, x \in \mathbb{R}^d)$ *be defined by (4.20) and assume (4.23).* *Then*

$$S^{N,\frac{1}{\gamma}}_{[A_1,A_2]}(u(\cdot, x)) \to_{N\to\infty} \mathbb{E} |U(1, 0)|^{\frac{1}{\gamma}} (A_2 - A_1) \text{ in probability}$$

where U *is given by (4.27).*

Proof In a first step, we will show that the p-variation of the random field Y given by (4.28) vanishes, for every $p \geq \frac{1}{\gamma}$. Indeed, for $p \geq \frac{1}{\gamma} > 1$

$$\sum_{i=0}^{N-1} |Y(t_{i+1}, x) - Y(t_i, x)|^p$$

$$\leq \sup_{|a-b|\leq \frac{A_2-A_1}{N}} |Y(a, x) - Y(b, x)|^{p-1} \sum_{i=0}^{N-1} |Y(t_{i+1}, x) - Y(t_i, x)|.$$

Recall from Proposition 4.30 that Y has absolute continuous temporal sample paths. The continuity of Y with respect to the time variable and $p > 1$ implies that

$$\sup_{|a-b|\leq \frac{A_2-A_1}{N}} |Y(a, x) - Y(b, x)|^{p-1} \to_{N\to\infty} 0$$

pointwise while the quantity $\sum_{i=0}^{N-1} |Y(t_{i+1}, x) - Y(t_i, x)|$ is bounded by the total variation of $t \to Y(t, x)$ over the interval $[A_1, A_2]$. Consequently,

$$\sum_{i=0}^{N-1} |Y(t_{i+1}, x) - Y(t_i, x)|^p \to_{N\to\infty} 0 \text{ pointwise.} \tag{4.32}$$

Let us now analyze the $\frac{1}{\gamma}$-variation of the random field U. We have, by the scaling property in time of U obtained in Proposition 4.29, if "$=^{(d)}$" stands for the equality in distribution,

$$\sum_{i=0}^{N-1} |U(t_{i+1}, x) - U(t_i, x)|^p =^{(d)} (A_2 - A_1)^{\gamma p} \frac{1}{n^{\gamma p}} \sum_{i=0}^{N-1} |U(i + 1, x) - U(i, x)|^p$$

$$= (A_2 - A_1)^{\gamma p} \frac{1}{n^{\gamma p-1}} V_N \tag{4.33}$$

with

$$V_N = \frac{1}{N} \sum_{i=0}^{N-1} |U(i+1,x) - U(i,x)|^p .$$

The sequence $(U(i+1,x) - U(i,x), i \geq 0)$ is stationary due to the fact that $(U(t,x), t \geq 0)$ has stationary increments, see Proposition 4.29. On the other hand, $U(t,x)$ is an element of the qth Wiener chaos, for every $t \geq 0$, $x \in \mathbb{R}^d$. Actually, we have

$$U(t,x) = \int_{\mathbb{R}} \int_{\mathbb{R}^d} G_{t,x}\left((s_1, z_1), \ldots, (s_q, z_q)\right) W(ds_1, dz_1) \ldots W(ds_q, dz_q)$$

where

$$G_{t,x}\left((s_1, z_1), \ldots (s_q, z_q)\right)$$
$$= c(\mathbf{H}, q) \int_{\mathbb{R}} du \int_{\mathbb{R}^d} dy \left(G((t-u)_+, x-y) - G((-u)_+, x-y)\right)$$
$$\times (u - s_1)_+^{-\left(\frac{1}{2} + \frac{1-H_0}{q}\right)} \ldots (u - s_q)_+^{-\left(\frac{1}{2} + \frac{1-H_0}{q}\right)}$$
$$\times (y - z_1)_+^{-\left(\frac{1}{2} + \frac{1-\mathbf{H}}{q}\right)} \ldots (y - z_q)_+^{-\left(\frac{1}{2} + \frac{1-\mathbf{H}}{q}\right)} .$$

Moreover,

$$G_{t,x}\left((s_1, z_1), \ldots (s_q, z_q)\right) - G_{s,x}\left((s_1, z_1), \ldots (s_q, z_q)\right)$$
$$= c(\mathbf{H}, q) \int_{\mathbb{R}} du \int_{\mathbb{R}^d} dy \left(G((t-u)_+, x-y) - G((s-u)_+, x-y)\right)$$
$$\times (u - s_1)_+^{-\left(\frac{1}{2} + \frac{1-H_0}{q}\right)} \ldots (u - s_q)_+^{-\left(\frac{1}{2} + \frac{1-H_0}{q}\right)}$$
$$\times (y - z_1)_+^{-\left(\frac{1}{2} + \frac{1-\mathbf{H}}{q}\right)} \ldots (y - z_q)_+^{-\left(\frac{1}{2} + \frac{1-\mathbf{H}}{q}\right)}$$
$$= c(\mathbf{H}, q) \int_{\mathbb{R}} du \int_{\mathbb{R}^d} dy \left(G((t-s-u)_+, x-y) - G((-u)_+, x-y)\right)$$
$$\times (u - (s_1 - s))_+^{-\left(\frac{1}{2} + \frac{1-H_0}{q}\right)} \ldots (u - (s_q - s))_+^{-\left(\frac{1}{2} + \frac{1-H_0}{q}\right)}$$
$$\times (y - z_1)_+^{-\left(\frac{1}{2} + \frac{1-\mathbf{H}}{q}\right)} \ldots (y - z_q)_+^{-\left(\frac{1}{2} + \frac{1-\mathbf{H}}{q}\right)}$$

so

$$G_{t,x}\left((s_1, z_1), \ldots (s_q, z_q)\right) - G_{s,x}\left((s_1, z_1), \ldots (s_q, z_q)\right)$$
$$= G_{t-s,x}\left((s_1 - s, z_1), \ldots (s_q - s, z_q)\right). \tag{4.34}$$

Since the kernel $G_{t,x}$ satisfies the shifting property (4.34), it follows from Theorem 8.3.1 in [37] that the sequence $(U(i+1,x) - U(i,x), i \geq 0)$ is also mixing. Therefore (see e.g. Chap. 2 in [37])

$$V_N \to_{N\to\infty} \mathbf{E}\,|U(1,0)|^p \text{ almost surely and in } L^1(\Omega). \tag{4.35}$$

By (4.33) and (4.35) we obtain (since the convergence in law to a constant implies the convergence in probability)

$$\sum_{i=0}^{N-1} |U(t_{i+1},x) - U(t_i,x)|^p \to_{N\to\infty} \begin{cases} 0, & \text{if } p > \frac{1}{\gamma} \\ \mathbf{E}\,|U(1,0)|^{\frac{1}{\gamma}}(A_2 - A_1) & \text{if } p = \frac{1}{\gamma} \\ +\infty & \text{if } p < \frac{1}{\gamma} \end{cases} \tag{4.36}$$

in probability. Now, by using Minkovski's inequality

$$\left(\sum_{i=0}^{N-1} |U(t_{i+1},x) - U(t_i,x)|^p \right)^{\frac{1}{p}} - \left(\sum_{i=0}^{N-1} |Y(t_{i+1},x) - Y(t_i,x)|^p \right)^{\frac{1}{p}}$$

$$\leq S_{[A_1,A_2]}^{N,p}(u(\cdot,x)))$$

$$\leq \left(\sum_{i=0}^{N-1} |U(t_{i+1},x) - U(t_i,x)|^p \right)^{\frac{1}{p}} + \left(\sum_{i=0}^{N-1} |Y(t_{i+1},x) - Y(t_i,x)|^p \right)^{\frac{1}{p}}.$$

To get the conclusion, it suffices to use (4.32) and (4.36). ∎

Remark 4.8 From the proof of Theorem 4.5, we can notice that the solution (4.20) has zero p-variation in time on any interval $[A_1, A_2]$ if $p > \frac{1}{\gamma}$.

Chapter 5
Statistical Inference for Stochastic (Partial) Differential Equations with Hermite Noise

While the statistical inference for SDEs and SPDEs driven by the Brownian motion or, more general, by a Gaussian noise, has a long history, the statistical inference for systems driven by Hermite processes and sheets is at its beginning. Our aim is to illustrate how certain parameters can be estimated in two situations where the Hermite processes appear as random perturbations.

The first example concerns the Hermite Ornstein-Uhlenbeck process introduced and analyzed in Sect. 3.4. We recal that this stochastic process is the solution to the Langevin equation with Hermite noise (3.13). In (3.13), we will assume that the initial value ξ vanishes (for simplicity) and the drift coefficient is $\lambda = 1$ (the estimation of this parameter has been done in [28]), so our observed process solves the equation

$$X_t = - \int_0^t X_s ds + \sigma Z_t^{H,q}, t \geq 0. \tag{5.1}$$

The purpose is to estimate the remaining two parameters, i.e. the Hurst index H and the volatility coefficient σ, from the onservation of the solution to (5.1) at the discrete times $t_i = \frac{i}{N}, i - 0, 1, ..., N$ along the unit interval $[0, 1]$. Our estimators will be expressed in terms of the quadratic (and generalized) variation of the Hermite Ornstein-Uhlenbeck process. Therefore, we need to analyze the asymptotic behavior of these random sequences, i.e. the sequences $V_N(X)$ given by (5.7) and the $\frac{1}{H}$-variation defined by the left-hand side of (5.31). To this end, we notice that the solution to the Langevin equation (5.1) can be written as the sum of the noise term (the Hermite process) and of another more regular stochastic process, and we show the asymptotic behavior of the power variation of the solution is given by the power variation of the noise term, which has already been studied in Chap. 2. Then, we construct in a standard way, estimators for the parameters H and σ and we deduce the asymptotic properties of these estimators. This has been essentially treated in [2].

© The Author(s), under exclusive license to Springer Nature Switzerland AG 2023
C. Tudor, *Non-Gaussian Selfsimilar Stochastic Processes*,
SpringerBriefs in Probability and Mathematical Statistics,
https://doi.org/10.1007/978-3-031-33772-7_5

In the second example we deal with the stochastic heat equation driven by the Hermite sheet from Sect. 4.4. Here we illustrate how the drift parameter θ appearing in the SPDE (5.33) can be estimated. The approach is also based on the analysis of the temporal p- variation of the mild solution u_θ to (5.33), i.e. the of the sequence

$$\sum_{i=0}^{N-1} |u_\theta(t_{i+1}, x) - u_\theta(t_i, x)|^p ,$$

where $t_i = \frac{i}{N}, i = 0, 1, ..., N$ and $x \in \mathbb{R}^d$ is fixed.

5.1 Parameter Identification for the Hermite Ornstein-Uhlenbeck Process

Let us consider the Hermite Ornstein-Uhlenbeck (HOU) process given by (3.14) with $\xi = 0$ and $\lambda = 1$. That is,

$$X_t = \sigma e^{-t} \int_0^t e^u dZ_u^{H,q} . \tag{5.2}$$

We can also write, due to Definition 3.1,

$$X_t = I_q(h_t),$$

where, for $t \geq 0$ and $y_1, ..., y_q \in \mathbb{R}$,

$$h_t(y_1, .., y_q) = \sigma c(H, q) \int_0^t e^{-(t-u)} \prod_{i=1}^{q} (u - y_i)_+^{-\left(\frac{1}{2} + \frac{1-H}{q}\right)} du, \tag{5.3}$$

where $c(H, q)$ is the constant given by (2.5). Via the Langevin equation (3.13), we have the decomposition, for $t \geq 0$,

$$X_t = -\int_0^t X_s ds + \sigma Z_t^{H,q} = Y_t + \sigma Z_t^{H,q} \tag{5.4}$$

with

$$Y_t = -\int_0^t X_s ds, \qquad t \geq 0. \tag{5.5}$$

5.1.1 Quadratic Variation

Our purpose is to use the p- variation of the process (5.2) in order to identify its Hurst and diffusion parameters. Let us consider the partition of the unit interval $[0, 1]$ given by

$$t_i = \frac{i}{N} \text{ for } i = 0, ..., N \text{ and for } N \geq 1 \tag{5.6}$$

We will assume that the process X is observed at times t_i and we will define estimators for the parameters H and σ in (5.4) based on these observations. Define, for every $N \geq 1$, the sequence of (centered and renormalized) quadratic variations

$$V_N(X) = \frac{1}{N} \sum_{i=0}^{N-1} \left[\frac{\left(X_{t_{i+1}} - X_{t_i}\right)^2}{\sigma^2 \mathbf{E}\left(Z_{t_{i+1}}^{H,q} - Z_{t_i}^{H,q}\right)^2} - 1 \right] \tag{5.7}$$

$$= \frac{1}{N} \sum_{i=0}^{N-1} \left[\frac{N^{2H}}{\sigma^2} \left(X_{t_{i+1}} - X_{t_i}\right)^2 - 1 \right].$$

We aim at finding the limit behavior, as $N \to \infty$, of the sequence $V_N(X)$. We will benefit from the behavior of the quadratic variation of the noise of (5.4) which is well-known: while for $q = 1$, this is the famous Breuer-Major theorem (see e.g. [8, 27]), for $q \geq 2$, it has been obtained in [11, 49]. Let us recall these results. By "$\to^{(d)}$" we denote the convergence in distribution and by $N(0, 1)$ we indicate the standard normal law.

Theorem 5.6 *Assume $H \in \left(\frac{1}{2}, 1\right)$ and $q \geq 1$ integer. Let $V_N(Z^{H,q})$ be given by*

$$V_N(Z^{H,q}) = \frac{1}{N} \sum_{i=0}^{N-1} \left[\frac{\left(Z_{t_{i+1}}^{H,q} - Z_{t_i}^{H,q}\right)^2}{\mathbf{E}\left(Z_{t_{i+1}}^{H,q} - Z_{t_i}^{H,q}\right)^2} - 1 \right]. \tag{5.8}$$

Then

1. If $q = 1$ and $H \in \left(\frac{1}{2}, \frac{3}{4}\right)$,

$$K_{1,1}\sqrt{N} V_N(Z^{H,1}) \to_{N\to\infty}^{(d)} N(0, 1). \tag{5.9}$$

2. If $q \geq 2$ or $q = 1$ and $H > \frac{3}{4}$, then

$$K_q N^{\frac{2-2H}{q}} V_N(Z^{H,q}) \to_{N\to\infty} Z_1^{H',2} \text{ in } L^2(\Omega) \tag{5.10}$$

where $Z_1^{H',2}$ is a Rosenblatt random variable with Hurst parameter $H' = \frac{2(H-1)}{q} + 1$. The constants $K_{1,1}, K_i, i = 1, .., q$ are explicit, we refer to [11, 27] for their expression.

Actually the result at point 1. above holds for every $H \in (0, \frac{3}{4})$ (and even for $H = \frac{3}{4}$ under a different renormalization) but this case will be not discussed here.

We will prove in the sequel that $V_N(X)$ keeps the same behavior as the quadratic variations of the noise of (5.4). This means that the drift process $(Y_t, t \in [0, T])$ given by (5.5) does not affect the behavior of $V_N(X)$. This is due to the regularity of Y and to the fact that the correlation between the increments of Y and $Z^{H,q}$ is weak enough.

Proposition 5.31 *Let $V_N(X)$ be given by (5.7).*

1. *Assume $H \in \left(\frac{1}{2}, 1\right)$ and $q \geq 2$. Then, with $K_q, Z_1^{H',2}$ from (5.10)*

$$K_q N^{\frac{2-2H}{q}} V_N(X) \to_{N\to\infty} Z_1^{H',2} \text{ in } L^2(\Omega).$$

2. *If $q = 1$ and $H \in \left(\frac{1}{2}, \frac{3}{4}\right)$,*

$$K_{1,1}\sqrt{N} V_N(X) \to_{N\to\infty}^{(d)} N(0, 1).$$

with $K_{1,1}$ from (5.9).

Proof Assume first $q \geq 2$. We decompose $V_N(X)$ as follows:

$$V_N(X) = \frac{1}{N} \sum_{i=0}^{N-1} \left[N^{2H} \left(Z_{t_{i+1}}^{H,q} - Z_{t_i}^{H,q} \right)^2 - 1 \right] + \frac{1}{\sigma^2} \frac{1}{N} \sum_{i=0}^{N-1} N^{2H} \left(Y_{t_{i+1}} - Y_{t_i} \right)^2$$

$$+ \frac{2}{\sigma} \frac{1}{N} \sum_{i=0}^{N-1} N^{2H} \left(Y_{t_{i+1}} - Y_{t_i} \right) \left(Z_{t_{i+1}}^{H,q} - Z_{t_i}^{(q,H)} \right)$$

$$= V_N(Z^{(q,H)}) + T_{1,N} + T_{2,N} \tag{5.11}$$

with

$$T_{1,N} = \frac{1}{\sigma^2} N^{2H-1} \sum_{i=0}^{N-1} \left(Y_{t_{i+1}} - Y_{t_i} \right)^2 \tag{5.12}$$

and

$$T_{2,N} = \frac{2}{\sigma} N^{2H-1} \sum_{i=0}^{N-1} \left(Y_{t_{i+1}} - Y_{t_i} \right) \left(Z_{t_{i+1}}^{H,q} - Z_{t_i}^{H,q} \right). \tag{5.13}$$

In (5.11), the limit of the sequence $V_N(Z^{H,q})$ is known from Theorem 5.6. We will prove that the other terms does not contribute to the limit. That is, we show that, for $i = 1, 2$ and for every $p \geq 1$

$$N^{\frac{2-2H}{q}} T_{i,N} \to_{N\to\infty} 0 \text{ in } L^P(\Omega). \tag{5.14}$$

The summand $T_{1,N}$ can be easily estimated, by using Hölder's inequality. Indeed, for every $p \geq 1$,

$$\mathbf{E}|T_{1,N}|^p = \sigma^{-2p} N^{(2H-1)p} \mathbf{E} \left| \sum_{i=0}^{N-1} \left(\int_{t_i}^{t_{i+1}} X_s ds \right)^2 \right|^p \tag{5.15}$$

$$\leq \sigma^{-2p} N^{(2H-1)p} N^{(p-1)} \mathbf{E} \sum_{i=0}^{N-1} \left(\int_{t_i}^{t_{i+1}} X_s ds \right)^{2p}$$

$$\leq \sigma^{-2p} N^{(2H-1)p} N^{-p} \mathbf{E} \sum_{i=0}^{N-1} \int_{t_i}^{t_{i+1}} |X_s|^{2p} ds \leq CN^{(2H-2)p} \tag{5.16}$$

where we used (3.16). Consequently (5.14) holds true for $i = 1$, since for every $p \geq 1$

$$\mathbf{E} \left| N^{\frac{2-2H}{q}} T_{1,N} \right|^p \leq CN^{(2H-2)(1-\frac{1}{q})p} \to_{N\to\infty} 0.$$

To analyze the term $T_{2,N}$, we need to use its Wiener chaos decomposition (actually, a direct proof based on Hölder inequality can be done only for $q \geq 3$). We can write, for every $N \geq 1$ and for every $i = 0, ..., N$,

$$Y_{t_{i+1}} - Y_{t_i} = I_q(h_{i,N})$$

where $h_{i,N} = h_{t_{i+1}} - h_{t_i}$ (h_t is given in (5.3)), i.e.

$$h_{i,N}(y_1, \ldots, y_q) = \sigma d(q, H) \int_{t_i}^{t_{i+1}} ds \int_0^s du e^{-(s-u)} \prod_{l=1}^q (u - y_l)_+^{-(\frac{1}{2} + \frac{1-H}{q})} \tag{5.17}$$

and

$$Z_{t_{i+1}}^{H,q} - Z_{t_i}^{H,q} = I_q(\ell_{i,N})$$

with (recall that $L_t = L^{H,q}$ is the kernel of the Hermite process, see (2.4))

$$\ell_{i,N}(y_1, \ldots, y_q) = L_{t_{i+1}}(y_1, \ldots, y_q) - L_{t_i}(y_1, \ldots, y_q) \tag{5.18}$$

for every $y_1, ..., y_q \in \mathbb{R}$. In this way, via the product formula for multiple integrals (1.33),

$$T_{2,N} = \frac{2}{\sigma} N^{2H-1} \sum_{i=0}^{N-1} I_q(h_{i,N}) I_q(\ell_{i,N})$$

$$= \frac{2}{\sigma} N^{2H-1} \sum_{i=0}^{N-1} \sum_{r=0}^{q} r! (C_q^r)^2 I_{2q-2r}(h_{i,N} \otimes_r \ell_{i,N}) := \sum_{r=0}^{q} T_{2,N}^{(r)}$$

with, for $r = 0, .., q$,

$$T_{2,N}^{(r)} = \frac{2}{\sigma} N^{2H-1} r! \binom{r}{q}^2 \sum_{i=0}^{N-1} I_{2q-2r}(h_{i,N} \otimes_r \ell_{i,N}). \tag{5.19}$$

We will obtain (5.14) for $i = 2$ if we show that

$$\mathbf{E}|N^{\frac{2-2H}{q}} T_{2,N}^{(r)}|^2 \to_{N\to\infty} 0 \tag{5.20}$$

via the hypercontractivity property (1.31). Assume $r = q$. Notice that $h_{i,N}, \ell_{i,N}$ are symmetric functions. Below, we denote by c, C generic strictly positive constants not depending on N and that are allowed to change from one line to another. We have

$$T_{2,N}^{(q)} = cN^{2H-1} \sum_{i=0}^{N-1} \langle h_{i,N}, \ell_{i,N} \rangle_{L^2(\mathbb{R}^q)}$$

with $h_{i,N}, \ell_{i,N}$ given by (5.17), (5.18) respectively. Thus $T_{2,N}^{(q)}$ is a deterministic sequence and

$$|T_{2,N}^{(q)}| = cN^{2H-1} \sum_{i=0}^{N-1} \int_{\mathbb{R}^q} dy_1 ... dy_q \int_{t_i}^{t_{i+1}} ds \int_0^s du e^{-(s-u)}$$

$$\times \prod_{j=1}^{q} (u - y_j)_+^{-(\frac{1}{2}+\frac{1-H}{q})} \int_{t_i}^{t_{i+1}} dv \prod_{j=1}^{q} (v - y_j)_+^{-(\frac{1}{2}+\frac{1-H}{q})}.$$

By using the formula (2.3) in Lemma 2.2, we obtain

$$|T_{2,N}^{(q)}| = cN^{2H-1} \sum_{i=0}^{N-1} \int_{t_i}^{t_{i+1}} ds \int_0^s du e^{-(s-u)} \int_{t_i}^{t_{i+1}} dv |u - v|^{2H-2}$$

$$\leq cN^{2H-1} \sum_{i=0}^{N-1} \int_{t_i}^{t_{i+1}} ds \int_0^1 du \int_{t_i}^{t_{i+1}} dv |u - v|^{2H-2} \leq CN^{2H-2} \tag{5.21}$$

and (5.20) holds for $r = q$ because we assumed $q \geq 2$.

Now assume $1 \leq r \leq q - 1$. We study the sequence

$$T_{2,N}^{(r)} = cN^{2H-1}I_{2q-2r}\left(\sum_{i=0}^{N-1} h_{i,N} \otimes_r \ell_{i,N}\right)$$

where, again via (2.3) and Fubini,

$$(h_{i,N} \otimes_r \ell_{i,N})(y_1, ..., y_{2q-2r})$$

$$= cN^{2H-1}\sum_{i=0}^{N-1}\int_{t_i}^{t_{i+1}} ds \int_0^s du\, e^{-(s-u)}\int_{t_i}^{t_{i+1}} dv|u-v|^{(2H-2)\frac{r}{q}}$$

$$(u-y_1)_+^{-(\frac{1}{2}+\frac{1-H}{q})}...(u-y_{q-r})_+^{-(\frac{1}{2}+\frac{1-H}{q})}$$

$$\times (v-y_{q-r+1})_+^{-(\frac{1}{2}+\frac{1-H}{q})}...(v-y_{2q-2r})_+^{-(\frac{1}{2}+\frac{1-H}{q})}$$

for every $y_1, ..., y_{2q-2r} \in \mathbb{R}$. By isometry (see (3.6))

$$\mathbf{E}\left|T_{2,N}^{(r)}\right|^2 \tag{5.22}$$

$$= cN^{4H-2}\|\sum_{i=0}^{N-1} h_{i,N}\tilde{\otimes}_r \ell_{i,N}\|_{L^2(\mathbb{R}^{2q-2r})}^2 \le cN^{4H-2}\|\sum_{i=0}^{N-1} h_{i,N} \otimes_r \ell_{i,N}\|_{L^2(\mathbb{R}^{2q-2r})}^2$$

$$= CN^{4H-2}\sum_{i,j=0}^{N-1}\langle h_{i,N} \otimes_r \ell_{i,N}, h_{j,N} \otimes_r \ell_{j,N}\rangle_{L^2(\mathbb{R}^{2q-2r})}$$

$$= CN^{4H-2}\sum_{i,j=0}^{N-1}\int_{t_i}^{t_{i+1}} ds \int_0^s du\, e^{-(s-u)}\int_{t_i}^{t_{i+1}} dv \int_{t_j}^{t_{j+1}} ds' \int_0^{s'} du'\, e^{-(s'-u')}\int_{t_j}^{t_{j+1}} dv'$$

$$\times |u-v|^{(2H-2)\frac{r}{q}}|u'-v'|^{(2H-2)\frac{r}{q}}|u-u'|^{(2H-2)\frac{(q-r)}{q}}|v-v'|^{(2H-2)\frac{(q-r)}{q}} \tag{5.23}$$

where we used again (5.23). Now, we majorize the exponential function by 1 and the integral over $[0, s] \times [0, s']$ by the integral over $[0, 1]^2$. We will obtain, for every $r = 1, ..., q - 1$,

$$\mathbf{E}\left|T_{2,N}^{(r)}\right|^2 \le CN^{4H-2}\sum_{i,j=0}^{N-1}\int_0^1 du \int_0^1 du' \int_{t_i}^{t_{i+1}} dv \int_{t_j}^{t_{j+1}} dv'$$

$$\times |u-v|^{(2H-2)\frac{r}{q}}|u'-v'|^{(2H-2)\frac{r}{q}}|u-u'|^{(2H-2)\frac{(q-r)}{q}}|v-v'|^{(2H-2)\frac{(q-r)}{q}}$$

$$\le CN^{4H-4} \tag{5.24}$$

and consequently (5.20) holds since (see e.g. [11])

$$\int_{[0,1]^4} dudu'dvdv'|u-v|^{(2H-2)\frac{r}{q}}|u'-v'|^{(2H-2)\frac{r}{q}}|u-u'|^{(2H-2)\frac{(q-r)}{q}}|v-v'|^{(2H-2)\frac{(q-r)}{q}} < \infty$$

and since (5.24) and (5.21) give, for every $p \geq 2$

$$\mathbf{E}|T_{2,N}|^p \leq CN^{(2H-2)p}. \tag{5.25}$$

Assume $q = 1$ and $H \in \left(\frac{1}{2}, \frac{3}{4}\right)$. In this case we still have the decomposition (5.11) and the estimates (5.16), (5.21) and (5.24) (with $r = 0$). These bounds clearly imply, via (1.31), that for every $p \geq 1$

$$\mathbf{E}\left|\sqrt{N}T_{i,N}\right|^p \leq CN^{(2H-\frac{3}{2})p} \to_{N\to\infty} 0 \text{ for } i = 1, 2 \tag{5.26}$$

for $H < \frac{3}{4}$, with $T_{1,N}, T_{2,N}$ given by (5.12), (5.13) respectively.   ∎

Proposition 5.31 also shows (together with (1.31)) that the sequence $V_N(X)$ converges to zero in $L^p(\Omega)$ ($p \geq 1$) as $N \to \infty$. Via a Borel-Cantelli argument, we can easily obtain the almost sure convergence of $V_N(X)$ to zero as $N \to \infty$. This will be needed later for the estimation of the Hurst parameter.

Corollary 5.3 *Let $V_N(X)$ be given by (5.7). Then for every $q \geq 2$ (if $H \in \left(\frac{1}{2}, 1\right)$) and for $q = 1$ (if $H \in \left(\frac{1}{2}, \frac{3}{4}\right)$), the sequence $V_N(X)$ converges to zero almost surely as $N \to \infty$.*

Proof Let $q \geq 2$ and take $\gamma \in (0, 1 - H)$. Then for every $p \geq 1$, from Proposition 5.31

$$P(V_N(X) \geq N^{-\gamma}) \leq CN^{-\gamma p}\mathbf{E}|V_N|^p \leq cN^{p(H+\gamma-1)}.$$

By choosing p large enough we will have that $\sum_{N\geq 1} P(V_N(X) \geq N^{-\gamma})$ is convergent and the Borel-Cantelli lemma gives the conclusion.

For $q = 1$ and $H \in \left(0, \frac{3}{4}\right)$, we proceed similarly by taking $\gamma \in \left(0, \frac{1}{2}\right)$.   ∎

5.1.2   Estimation of the Hurst Parameter

The idea to construct an estimator for the Hurst parameter H in terms of the observation of the solution to (5.4) at times $X_{\frac{i}{N}}, i = 0, 1, \ldots, N$ is standard: one starts with an evaluation of $\mathbf{E}S_N$ with

$$S_N = \frac{1}{N}\sum_{i=0}^{N-1}\left(X_{t_{i+1}} - X_{t_i}\right)^2$$

yielding

$$\mathbf{E}S_N = \sigma^2 N^{-2H} + \sigma^2 N^{-2H}\left(\mathbf{E}T_{1,N} + \mathbf{E}T_{2,N}\right)$$

with $T_{1,N}, T_{2,N}$ given by (5.12), (5.13) respectively. From the proof of Proposition 5.31, the last two summands converge to zero as $N \to \infty$. Indeed, by using the estimates (5.16) and (5.25) we have for $i = 1, 2$,

$$N^{-2H} \mathbf{E} \left| T_{i,N} \right| \le N^{-2H} \left(\mathbf{E} \left| T_{i,N} \right|^2 \right)^{\frac{1}{2}} \le C N^{-2} \to_{N \to \infty} 0.$$

Consequently, we have (in the sequel $a_N \sim b_N$ means that the sequences a_N and b_N have the same behavior at infinity)

$$\mathbf{E} S_N \sim \sigma^2 N^{-2H}$$

and by taking the logarithm and by approximating $\mathbf{E} S_N$ by S_N

$$\log(S_N) \sim -2H \log(N) + \log(\sigma^2) \sim -2H \log(N). \tag{5.27}$$

This leads to a natural estimator

$$\widehat{H}_N = \frac{-\log(S_N)}{2 \log(N)}. \tag{5.28}$$

From Proposition 5.31 and Corollary 5.3 we immediately get the asymptotic behavior of the above estimator. The constants $K_{1,1}$, K_q and the random variable $Z_1^{H',2}$ are those from Proposition 5.31.

Proposition 5.32 *The estimator (5.28) is strongly consistent, i.e. $\widehat{H}_N$ converges almost surely to H as $N \to \infty$. Moreover, for $q \ge 2$*

$$K_q N^{\frac{2-2H}{q}} \left[2 \log(N) \left(H - \widehat{H}_N \right) - \log(\sigma^2) \right] \to_{N \to \infty}^{(d)} Z_1^{(2, H')}$$

while for $q = 1$ and $H \in \left(\frac{1}{2}, \frac{3}{4} \right)$,

$$K_{1,1} \sqrt{N} \left[2 \log(N) \left(H - \widehat{H}_N \right) - \log(\sigma^2) \right] \to_{N \to \infty}^{(d)} N(0, 1).$$

Proof For $N \ge 1$, we have

$$1 + V_N(X) = \frac{N^{2H}}{\sigma^2} S_N$$

and by taking the logarithm and using Corollary 5.3, almost surely,

$$V_N(X) \sim \log(1 + V_N(X)) = 2H \log(N) + \log(S_N) - \log(\sigma^2). \tag{5.29}$$

By combining (5.29) with (5.28), we can write for N large, almost surely,

$$H - \widehat{H}_N \sim \frac{V_N(X) + \log(\sigma^2)}{2 \log N}. \tag{5.30}$$

The strong consistency of $\widehat{H}_N$ (that is, $H - \widehat{H}_N$ converges almost surely to zero as $N \to \infty$) follows from (5.30) and Corollary 5.3 while the limit distribution is obtained from Proposition 5.31. ∎

Remark 5.9 The estimator (5.28) does not depend on σ so we can assume that this coefficient is unknown. On the other hand, when σ is known, we can also use relation (5.27) to define $\widetilde{H}_N = \frac{\log(\sigma^2) - \log(S_N)}{2\log(N)}$. This estimator will satisfy the limit theorem (when $q \geq 2$)

$$2K_q \log(N) N^{\frac{2-2H}{q}} \left(H - \widetilde{H}_N \right) \to_{N\to\infty}^{(d)} Z_1^{H',2}.$$

5.1.3　Estimation of σ

In this last section, let us discuss the estimation of the diffusion parameter σ by assuming that H is known. Actually it can also be estimated from the discrete observation of the process X, via generalized variations.

Let us first recall the following result concerning the variations of the Hermite process (see Proposition 2.3).

Proposition 5.33 *If $Z^{H,q}$ is a Hermite process of order $q \geq 1$, then*

$$\sum_{i=0}^{N-1} \left| Z_{t_{i+1}}^{H,q} - Z_{t_i}^{H,q} \right|^{\frac{1}{H}} \to_{N\to\infty} \mathbf{E} \left| Z_1^{H,q} \right|^{\frac{1}{H}} \text{ in probability.}$$

The previous property can be easily transferred to the solution to (5.4).

Proposition 5.34 *Let X be the solution to (5.4). Then*

$$\sum_{i=0}^{N-1} \left| X_{t_{i+1}} - X_{t_i} \right|^{\frac{1}{H}} \to_{N\to\infty} \sigma^{\frac{1}{H}} \mathbf{E} \left| Z_1^{H,q} \right|^{\frac{1}{H}} \text{ in probability.} \quad (5.31)$$

Proof By Minkovski's inequality,

$$\sigma \left(\sum_{i=0}^{N-1} \left| Z_{t_{i+1}}^{H,q} - Z_{t_i}^{H,q} \right|^{\frac{1}{H}} \right)^{H} - \left(\sum_{i=0}^{N-1} \left| Y_{t_{i+1}} - Y_{t_i} \right|^{\frac{1}{H}} \right)^{H} \leq \left(\sum_{i=0}^{N-1} \left| X_{t_{i+1}} - X_{t_i} \right|^{\frac{1}{H}} \right)^{H}$$

$$\leq \left(\sum_{i=0}^{N-1} \left| Y_{t_{i+1}} - Y_{t_i} \right|^{\frac{1}{H}} \right)^{H} + \sigma \left(\sum_{i=0}^{N-1} \left| Z_{t_{i+1}}^{H,q} - Z_{t_i}^{H,q} \right|^{\frac{1}{H}} \right)^{H}.$$

On the other hand,

$$\mathbf{E} \sum_{i=0}^{N-1} |Y_{t_{i+1}} - Y_{t_i}|^{\frac{1}{H}} \leq N^{1-\frac{1}{H}} \sum_{i=0}^{N-1} \int_{t_i}^{t_{i+1}} \mathbf{E}|X_s|^{\frac{1}{H}} \leq CN^{1-\frac{1}{H}}.$$

Consequently, $\left(\sum_{i=0}^{N-1} |Y_{t_{i+1}} - Y_{t_i}|^{\frac{1}{H}}\right)^H$ converges to zero in $L^1(\Omega)$ as $N \to \infty$ and the conclusion follows.   ∎

Consequently, define the estimator $\widehat{\sigma}_N$

$$\widehat{\sigma}_N := m(q, H)^{-H} \left(\sum_{i=0}^{N-1} |X_{t_{i+1}} - X_{t_i}|^{\frac{1}{H}}\right)^H \tag{5.32}$$

with $m(q, H) = \mathbf{E}\left|Z_1^{H,q}\right|^{\frac{1}{H}}$. Clearly, by Proposition 5.34

$$\widehat{\sigma}_N \to_{N\to\infty} \sigma$$

in probability, so $\widehat{\sigma}_N$ is a consistent estimator for σ.

5.2 Drift Estimation for the Stochastic Heat Equation with Hermite Noise

Let us now consider the parametrized version of the SPDE (4.17)

$$\frac{\partial u_\theta}{\partial t}(t, x) = \theta \Delta u_\theta(t, x) + \dot{Z}_{t,x}^{(H_0,\mathbf{H}),q,d+1}, \quad t \geq 0, x \in \mathbb{R}^d \tag{5.33}$$

with vanishing initial condition $u_\theta(0, x) = 0$ for every $x \in \mathbb{R}^d$. Our purpose is to estimate the drift parameter $\theta > 0$ in (5.33) based on the observation of the solution u_θ at discrete times and at a fixed point in space. That is, we assume that we have at our disposal the observations $(u_\theta(t_i, x), i = 0, 1, ..., N)$ with $t_i = \frac{i}{N}$ for $i = 0, 1, ..., N$ and $x \in \mathbb{R}^d$. Again, the noise $Z^{(H_0,\mathbf{H}),q,d+1}$ is a $d+1$- parameter Hermite process with Hurst index $(H_0, \mathbf{H}) \in \left(\frac{1}{2}, 1\right)^{d+1}$ where $\mathbf{H} = (H_1, \ldots, H_d)$.

Since the Green kernel that solves $\frac{\partial u_\theta}{\partial t}(t, x) = -\theta \Delta u_\theta(t, x)$ is $G(\theta t, x)$ with G defined by (4.18), the mild solution to (5.33) can be written as

$$u_\theta(t, x) = \int_0^t \int_{\mathbb{R}^d} G(\theta(t - s), x - y) dZ_{s,y}^{(H_0,\mathbf{H}),q,d+1} \tag{5.34}$$

where the integral $dZ_{s,y}^{(H_0,\mathbf{H}),q,d+1}$ is a Wiener-Hermite integral described in Section 4.3. It follows from Proposition 4.26 that (5.34) is well-defined if and only if condition (4.23) holds true. This will be assumed in what follows.

The key observation is that, via a trivial transform of the solution, we can move the drift θ in front of the random noise. This remark will allow to apply the p-variation method in order to estimate the drift parameter. This idea has been used in [24] or [34].

Let us define, for $t \geq 0$ and $x \in \mathbb{R}^d$,

$$v_\theta(t, x) = u_\theta\left(\frac{t}{\theta}, x\right). \tag{5.35}$$

We have the following result for v_θ.

Proposition 5.35 *The random field* $(v_\theta(t, x), t \geq 0, x \in \mathbb{R}^d)$ *satisfies, in the mild sense, the following SPDE*

$$\frac{\partial v_\theta}{\partial t} = \Delta v_\theta(t, x) + \theta^{-H_0}\dot{\tilde{Z}}_{t,x}^{(H_0,\mathbf{H}),q,d+1}, \quad t \geq 0, x \in \mathbb{R}^d \tag{5.36}$$

with vanishing initial condition $v_\theta(0, x) = 0$ *for every* $x \in \mathbb{R}^d$, *where* $\tilde{Z}^{(H_0,\mathbf{H}),q,d+1}$ *is an independent copy of* $Z^{(H_0,\mathbf{H}),q,d+1}$.

Proof We have, by the definition of the mild solution (5.34), for $t \geq 0$ and $x \in \mathbb{R}^d$,

$$v_\theta(t, x) = \int_0^{\frac{t}{\theta}} \int_{\mathbb{R}^d} G(t - \theta s, x - y)dZ_{s,y}^{(H_0,\mathbf{H}),q,d+1}$$

$$= \int_0^t \int_{\mathbb{R}^d} G(t - s, x - y)dZ_{\frac{s}{\theta},y}^{(H_0,\mathbf{H}),q,d+1}$$

and by using the H_0-self-similarity in time of $Z^{(H_0,\mathbf{H}),q,d+1}$, we obtain

$$v_\theta(t, x) \stackrel{(d)}{\equiv} \theta^{-H_0} \int_0^t \int_{\mathbb{R}^d} G(t - s, x - y)dZ_{s,y}^{(H_0,\mathbf{H}),q,d+1}$$

$$= \theta^{-H_0}u_1(t, x). \tag{5.37}$$

which means that v_θ verifies (5.36) in the mild sense. ∎

From Theorem 4.5, we deduce immediately the behavior of the p-variation in time for the random field u_θ. Recall the notation (4.31), for $p > 0, x \in \mathbb{R}^d$,

$$S_{[A_1,A_2]}^{N,p}(u_\theta(\cdot, x)) = \sum_{i=0}^{N-1} |u_\theta(t_{i+1}, x) - u_\theta(t_i, x)|^p$$

with $t_i, i = 0, 1, ..., N$ given by (4.30). If $[A_1, A_2] = [0, t]$, then we use the notation $S_{[A_1,A_2]}^{N,p} = S_t^{N,p}$. Recall that U is the pinned string process defined by (4.27).

Proposition 5.36 *Let u_θ be the solution to the parametrized heat equation defined by (5.34) and assume (4.23). Then*

$$S^{N,\frac{1}{\gamma}}_{[A_1,A_2]}(u_\theta(\cdot,x)) \to_{N\to\infty} \theta^{1-\frac{H_0}{\gamma}}(A_2-A_1)\mathbf{E}\,|U(1,0)|^{\frac{1}{\gamma}} \text{ in probability.}$$

Proof We have, for every $x \in \mathbb{R}^d$ and $N \geq 1$, via (5.35),

$$
\begin{aligned}
S^{N,\frac{1}{\gamma}}_{[A_1,A_2]}(u_\theta(\cdot,x)) &= \sum_{i=0}^{N-1} |u_\theta(t_{i+1},x) - u_\theta(t_i,x)|^{\frac{1}{\gamma}} \\
&= \sum_{i=0}^{N-1} |v_\theta(\theta t_{i+1},x) - v_\theta(\theta t_i,x)|^{\frac{1}{\gamma}} \\
&\stackrel{(d)}{=} \theta^{-\frac{H_0}{\gamma}} \sum_{i=0}^{N-1} |u_1(\theta t_{i+1},x) - u_1(\theta t_i,x)|^{\frac{1}{\gamma}}
\end{aligned}
$$

where for the last equality we used (5.37). Notice that $\theta t_i, i = 0,\ldots, N$ constitutes a partition of the interval $[\theta A_1, \theta A_2]$. Then, by Theorem 4.5,

$$\sum_{i=0}^{N-1} |u_1(\theta t_{i+1},x) - u_1(\theta t_i,x)|^{\frac{1}{\gamma}} \to_{N\to\infty} (\theta A_2 - \theta A_1)\mathbf{E}\,|U(1,0)|^{\frac{1}{\gamma}}.$$

Consequently,

$$S^{N,\frac{1}{\gamma}}_{[A_1,A_2]}(u_\theta(\cdot,x)) \to_{N\to\infty} \theta^{1-\frac{H_0}{\gamma}}(A_2-A_1)\mathbf{E}\,|U(1,0)|^{\frac{1}{\gamma}} \text{ in probability}$$

with U given by (4.27). ∎

Assume that we have the observations $(u_\theta(t_i,x), i = 0.1, .., N)$ with t_i given by (4.30), i.e. the solution is discretely observed during the time interval $[A_1, A_2]$ at some fixed point in space. From Proposition 5.36, we can define the following natural estimator for the drift parameter θ

$$\widehat{\theta}_N = \left(\frac{1}{(A_2-A_1)\mathbf{E}\,|U(1,0)|^{\frac{1}{\gamma}}} S^{N,\frac{1}{\gamma}}_{[A_1,A_2]}(u_\theta(\cdot,x)) \right)^{\frac{\gamma}{\gamma-H_0}}. \tag{5.38}$$

We deduce the following asymptotic behavior for (5.38).

Proposition 5.37 *The estimator $\widehat{\theta}_N$ given by (5.38) is consistent, i.e. $\widehat{\theta}_N$ converges in probability to θ as $N \to \infty$.*

Proof It is an immediate consequence of Proposition 5.36. ∎

Bibliography

1. Abry, P., Pipiras, V.: Wavelet based synthesis of the Rosenblatt process. Signal Process. **86**, 2326–2339 (2006)
2. Assaad, O., Tudor, C.A.: Pathwise analysis and parameter estimation for the stochastic Burgers equation. Bull. Sci. Math. **170**, 26, Paper No. 102995 (2021)
3. Assaad, O., Diez, C.-P., Tudor, C.A.: Generalized Wiener-Hermite integrals and rough non-Gaussian Ornstein-Uhlenbeck process. Stochastics **95**(2), 191–210 (2023)
4. Ayache, A., Esmili, Y.: Wavelet-type expansion of the generalized Rosenblatt process and its rate of convergence. J. Fourier Anal. Appl. **26**(3), 35, Paper No. 51 (2020)
5. Ayache, A., Hamonier, J., Loosveldt L.: Wavelet-Type Expansion of Generalized Hermite Processes with rate of convergence. Preprint (2023)
6. Ayache, A., Leger, S., Pontier, M.: Drap Brownien fractionnaire. Potential Anal. **17**(1), 31–43 (2002)
7. Bai, S., Taqqu, M.: Generalized Hermite processes, discrete chaos and limit theorems. Stoch. Process. Appl. **124**(4), 1710–1739 (2014)
8. Breuer, P., Major, P.: Central limit theorem for non-linear functionals of Gaussian fields. J. Mult. Anal. **13**, 425–441 (1983)
9. Chaurasia, A., Sehgal, V.K.: Synthetic traffic generation for wavelet based Rosenblatt process using LSTM for multicore architecture. Int. J. Control Autom. **12**(5), 607–617 (2019)
10. Cheridito, P., Kawaguchi, H., Maejima, M.: Fractional Ornstein-Uhlenbeck processes. Electron. J. Probab. **8**, 1–14 (2003)
11. Chronopoulou, A., Tudor, C.A., Viens, F.: Self-similarity parameter estimation and reproduction property for non-Gaussian Hermite processes. Commun. Stoch. Anal. **5**(1), 161–185 (2011)
12. Clarke De la Cerda, J., Tudor, C.A.: Wiener integrals with respect to the Hermite random field and applications to the wave equation. Collect. Math. **65**(3), 341–356 (2014)
13. Coupek, P., Duncan, T.E., Pasik-Duncan, B.: A Stochastic Calculus for Rosenblatt Processes. Preprint (2020)
14. Dalang, R.C.: Extending the martingale measure stochastic integral with applications to spatially homogeneous SPDE's. Electr. J. Probab. **4**, 1–29 (1999). (Erratum in Electr. J. Probab. **6**, 5 , 2001)

C. Tudor, *Non-Gaussian Selfsimilar Stochastic Processes*,
SpringerBriefs in Probability and Mathematical Statistics,
https://doi.org/10.1007/978-3-031-33772-7

15. Dieker, T.: Simulation of Fractional Brownian Motion. University of Twente, Amsterdam, M.Sc Theses (2004)
16. Diu Tran T.T.: Non-central limit theorems for quadratic functionals of Hermite-driven long memory moving average processes. Stoch. Dyn. **18**(4), 18, 1850028 (2018)
17. Dobrushin, R.L., Major, P.: Non-central limit theorems for non-linear functionals of Gaussian fields. Z. Wahrscheinlichkeitstheorie verw. Gebiete, 50, 27–52 (1979)
18. Embrechts, P., Maejima, M.: Selfsimilar Processes. Princeton University Press (2002)
19. Fauth, A., Tudor, C.A.: Multifractal random walk driven by a Hermite process. In: Handbook of High-frequency Trading and Modeling in Finance, pp. 221–249. Wiley Handbook. Finance Engineering Economic, Wiley, Hoboken, NJ (2016)
20. Gatheral, J., Jaisson, T., Rosenbaum, M.: Volatility is rough. Quant. Financ. **18**(6) (2018)
21. Harnett, D., Nualart, D.: Decomposition and limit theorems for a class of self-similar Gaussian processes. Stochastic analysis and related topics, vol. 72, pp. 99–116, Progress in Probability. Birkhäuser/Springer, Cham (2016)
22. Maejima, M., Tudor, C.A.: Wiener integrals with respect to the Hermite process and a non-central limit theorem. Stoch. Anal. Appl. **25**(5), 1043–1056 (2007)
23. Maejima, M., Tudor, C.A.: Selfsimilar processes with stationary increments in the second Wiener chaos. Probab. Math. Statist. **32**(1), 167–186 (2012)
24. Mahdi, Z., Tudor, C.A.: Estimation of the drift parameter for the fractional stochastic heat equation via power variation. Mod. Stoch. Theory Appl. **6**(4), 397–417 (2019)
25. Mueller, C., Tribe, R.: Hitting probabilities of a random string. Electron. J. Probab. **7**, 29, Paper No. 10 (2002)
26. Mueller, C., Wu, Z.: A connection between the stochastic heat equation and fractional Brownian motion, and a simple proof of a result of Talagrand. Electron. Commun. Probab. **f 14**(6), 55–65 (2009)
27. Nourdin, I., Peccati, G.: Normal Approximations with Malliavin Calculus From Stein's Method to Universality. Cambridge University Press (2012)
28. Nourdin, I., Diu Tran, T.T.: Statistical inference for Vasicek-type model driven by Hermite processes. Stoch. Process. Appl. **129**(10), 3774–3791 (2019)
29. Nualart, D.: Malliavin Calculus and Related Topics, 2nd edn. Springer (2006)
30. Pipiras, V., Taqqu, M.: Long-Range Dependence and Self-Similarity. Cambridge Series in Statistical and Probabilistic Mathematics. Cambridge University Press (2017)
31. Pipiras, V.: Wavelet-type expansion of the Rosenblatt process. J. Fourier Anal. Appl. **10**(6), 599–634 (2004)
32. Pipiras, V., Taqqu, M.S.: Integration questions related to fractional Brownian motion. Probab. Theory Related Fields **118**(2), 251–291 (2000)
33. Pipiras, V., Taqqu, M.: Regularization and integral representations of Hermite processes. Statist. Probab. Lett. **80**, 2014–2023 (2010)
34. Pospisil, J., Tribe, R.: Parameter estimates and exact variations for stochastic heat equations driven by space-time white noise. Stoch. Anal. Appl. **25**(3), 593–611 (2007)
35. Rosenblatt, M.: Independence and dependence. In: Proceedings of 4th Berkeley Symposium on Mathematical Statistics, vol. II, pp. 431–443 (1961)
36. Royden, H.L., Fitzpatrick, P.: Real Analysis, vol. 32. Macmillan New York (1988)
37. Samorodnitsky, G.: Stochastic Processes and Long Range Dependence. Springer Series in Operations Research and Financial Engineering, Springer, Cham (2016)
38. Slaoui, M., Tudor, C.A.: The linear stochastic heat equation with Hermite noise. Infin. Dimens. Anal. Quantum Probab. Relat. Top. **22**(3), 23, 1950022 (2020)
39. Slaoui, M., Tudor, C.A.: Behavior with respect to the Hurst index of the Wiener Hermite integrals and application to SPDEs. J. Math. Anal. Appl. **479**(1), 350–383 (2019)
40. Stoyanov, S., Rachev, S., Mittnik, S., Fabozzi, F.: Pricing derivatives in Hermite markets. Int. J. Theor. Appl. Finance **22**(6), 27, 1950031 (2019)
41. Taqqu, M.: Weak convergence to the fractional Brownian motion and to the Rosen-blatt process. Z. Wahrscheinlichkeitstheorie verw. Gebiete, 31, 287–302 (1975)
42. Taqqu, M.S.: A representation for self-similar process. Stoch. Process. Appl. **7**, 55–64 (1978)

43. Taqqu, M.S.: Convergence of integrated processes of arbitrary Hermite rank. Zeitschrift für Wahrscheinlichkeitstheorie und verwandte Gebiete **50**, 53–83 (1979)
44. Torres, S., Tudor, C.A.: Donsker type theorem for the Rosenblatt process and a binary market model. Stoch. Anal. Appl. **27**(3), 555–573 (2009)
45. Tudor, C.A., Xiao, Y.: Sample paths of the solution to the fractional-colored stochastic heat equation. Stoch. Dyn. **17**(1), 20, 1750004 (2017)
46. Tudor, C.A.: Analysis of variations for self-similar processes. In: A Stochastic Calculus Approach. Probability and its Applications (New York). Springer, Cham (2013)
47. Tudor, C.A.: Analysis of the Rosenblatt process. ESAIM Probab. Stat. **12**, 230–257 (2008)
48. Tudor, C.A.: Fractional stochastic heat equation with Hermite noise. Grad. J. Math. **7**(2), 1–18 (2022)
49. Tudor, C.A., Viens, F.: Variations and estimators through Malliavin calculus. Ann. Probab. **37**(6), 2093–2134 (2008)
50. Veillette, M., Taqqu, M.S.: Properties and numerical evaluation of the Rosenblatt distribution. Bernoulli **19**(3), 982–1005 (2013)
51. Walsh, J.B.: An introduction to stochastic partial differential equations. In: Ecole d'été de Probabilités de St. Flour XIV, vol. 1180, pp. 266-439. Lecture Notes in Mathematics. Springer, Berlin (1986)
52. Wheeden, R.L., Zygmund, A.: Measure and Integral. Marcel Dekker, New York-Basel (1977)

Printed in the United States
by Baker & Taylor Publisher Services